Drying of Biomass, Biosolids, and Coal

For Efficient Energy Supply and Environmental Benefits

Advances in Drying Science and Technology

Series Editor

Arun S. Mujumdar

McGill University, Quebec, Canada

For more information about this series, please visit: www.crcpress.com

Drying of Biomass, Biosolids, and Coal

For Efficient Energy Supply and Environmental Benefits

Edited by

Shusheng Pang, Sankar Bhattacharya, and Junjie Yan

CRC Press
Taylor & Francis Group
Boca Raton London New York

CRC Press is an imprint of the
Taylor & Francis Group, an **informa** business

CRC Press
Taylor & Francis Group
6000 Broken Sound Parkway NW, Suite 300
Boca Raton, FL 33487-2742

First issued in paperback 2020

ISBN-13: 978-1-138-54933-3 (hbk)
ISBN-13: 978-0-367-65650-8 (pbk)

Library of Congress Cataloging-in-Publication Data

Names: Pang, Shusheng, author. | Bhattacharya, Sankar (Chemical engineer) author. | Yan, Junjie (Power system engineer), author.
Title: Drying of biomass, biosolids, and coal: for efficient energy supply and environmental benefits / Shusheng Pang, Sankar Bhattacharya, Junjie Yan.
Description: First edition. | Boca Raton, FL : CRC Press/Taylor & Francis Group, 2018. | Series: Advancing in drying technology | Includes bibliographical references and index.
Identifiers: LCCN 2018049433| ISBN 9781138549333 (hardback : alk. paper) | ISBN 9781351000871 (ebook)
Subjects: LCSH: Biomass--Drying. | Coal--Drying. | Biomass conversion.
Classification: LCC TP339 .P354 2018 | DDC 662/.88--dc23
LC record available at https://lccn.loc.gov/2018049433

Visit the Taylor & Francis Web site at
http://www.taylorandfrancis.com

and the CRC Press Web site at
http://www.crcpress.com

Contents

Contents

Preface

It is well known that human beings are heavily reliant on fossil fuels (crude oil, natural gas, and coal) to meet the energy demand. The use of fossil fuels has caused concerns such as climate change and potential energy shortages. Therefore, finding alternative and sustainable energy resources has been a critical task in the world to secure the future energy supply and to reduce greenhouse gas emissions. In the meantime, the use of fossil fuel in an efficient and clean way will also help mitigate the above-mentioned issues. Biomass has been recognized as one of the most promising sources for future fuels and energy. The biomass, in the form of woodchips, sawdust, bagasse, grass, and agricultural residues, is generated from trees, agricultural crops, or purpose-planted coppices which absorb carbon dioxide through photosynthesis during plant growth. In this way, the whole system, from feedstock growing through energy processing to energy consumption, is carbon neutral.

Organic solid waste is another alternative source for energy, and effective use of this resource has double benefits for reducing landfill space and for recovery of energy. However, the characteristics of organic wastes vary significantly with respect to energy content, ash content, water content, species of microorganisms, and other potential contaminant elements. Sewage sludge is a typical organic solid waste stream from wastewater treatment plants. Use of the sewage sludge for energy faces challenges of high water content, while bound water behaves quite differently from free water.

Algae, as another alternative source for energy and fuel, covers many different organisms which grow in a similar manner as plants, using photosynthesis to convert solar energy and CO_2 into chemical energy in the form of oils, carbohydrates, and proteins. Benefits of using algae biofuel include high yield per crop area and CO_2 captured is high: up to two orders higher than terrestrial vegetations. Again, the high water content in fresh algae is one of the challenges in the processing of algae biomass to biodiesel and energy.

To increase the energy efficiency, improve the energy product quality, and reduce the emissions in the energy conversion in the utilization of biomass, organic solid wastes, and algae, drying of these feedstocks to required moisture content is important in the development of alternative energy systems. In a boiler, the energy efficiency can be increased by 5 to 15% with the increased combustion temperature and the reduced flue gas temperature by using the dried biomass instead of wet biomass and biosolids. In gasification and pyrolysis conversions, the feedstocks need to be dried to even lower moisture content from 5 to 10% to reduce the water content in the energy products.

However, there are challenges in the drying of these feedstocks, including low energy efficiency, negative environmental impacts, the risk of fire in drying, and high costs. To mitigate these issues, it is critical to understand the feedstock properties and drying process, to analyze the energy and exergy efficiencies, and to develop new drying technologies.

Due to the huge availability and low price, low-rank coals still make significant contribution to the world energy supply at present. However, extensive efforts have been made to process the low-rank coal more efficiently and cleanly. Drying of these coals is an important operation in the integrated system to achieve this goal.

This book is a collection of recent advances in the drying of biomass, sewage sludge, algae, and low-rank coals. It covers both fundamental studies on the drying process, and the development of new drying technologies. Important issues in the commercial drying operations are also tackled; these include energy and exergy efficiencies, environmental impact, and potential safety concerns. Also, the performance of energy production plants in integration with biomass drying has been assessed to provide information for plant optimization.

We are sure that this book will be useful to anyone who is interested in the utilization of biomass, organic solid wastes, algae, and low-rank coals for energy.

Dr. Shusheng Pang
Professor, Department of Chemical and Process Engineering,
University of Canterbury, Christchurch, New Zealand

Dr. Sankar Bhattacharya
Professor, Department of Chemical Engineering,
Monash University, Australia

Dr. Junjie Yan
Professor and Dean, School of Energy and Power Engineering,
Xi'an Jiaotong University, Xi'an, China

Editors

Shusheng Pang

Dr. Pang received his PhD from the Department of Chemical and Process Engineering, University of Canterbury in 1994. Then he worked at New Zealand Forest Research Institute for eight years on wood drying and processing of wood products. He was appointed associate professor at Canterbury University in 2002 and full professor in 2009. Dr. Pang is also the director of Canterbury University's Wood Technology Research Centre. He is a fellow of Engineering New Zealand and a fellow of the International Academy of Wood Science (IAWS). In addition, he serves as an editorial board member for *Energies*, *Biomass Conversion and Biorefinery*, and *Maderas. Ciencia y Tecnología*.

Dr. Pang's research interests and expertise include:

- Drying of wood, biomass, and bio-solid wastes;
- Processing of wood-based products;
- Biomass energy, biofuels and chemicals, and other clean energy technologies;
- Bio-composites and renewable chemicals/polymers from woody biomass.

In the last 25 years, Dr. Pang, with his team, has developed a number of advanced technologies and published over 150 papers in international leading journals and five book chapters in the above fields. Recently, Dr. Pang organized and co-chaired the 6th International Conference on Bio-refinery in Christchurch, New Zealand.

Sankar Bhattacharya

Dr. Sankar Bhattacharya has over 28 years' engineering and research experience in coal-fired power generation. Prior to his return to Australia in 2009, he was with the International Energy Agency (IEA) leading their Cleaner Fossil Fuels program, where he was responsible for IEA's work on coal-fired power stations and co-authored several publications including their flagship publications—World Energy Outlook, Energy Technology Perspectives, CO_2 Capture and Storage, and several country reviews. He previously worked in India on the design and commissioning services of coal-fired power stations, in Thailand on biomass carbonization, in Australia with the Lignite CRC as a principal research engineer, and with Anglo Coal as principal process engineer. He led the first pilot plant trials in Australia on circulating fluidized bed (CFB) combustion, and pressurized oxygen-blown gasification using Australian and North American lignites at facilities in Australia and the USA.

Dr. Bhattacharya is the co-author of two patents, six book chapters, over ten research reports, and over 100 journal publications. He currently leads a group of 18 researchers on coal- and biomass-based and industry-funded drying, gasification, oxy-fuel, CFB, and liquid fuels research projects at Monash University.

Junjie Yan

Dr. Yan is currently the Dean of the School of Energy and Power Engineering at Xi'an Jiaotong University, China. He received his PhD in Engineering from Xi'an Jiaotong University in 1994. He was a senior visiting scholar in the Department of Mechanical Engineering at Yokohama National University from September 2001 to March 2002.

Dr. Yan is a well-known researcher in the areas of thermal system energy saving, multiphase flow, and heat transfer. He is the author or co-author of several books and more than 300 peer-reviewed journal papers, about half of which are in international journals. He holds more than 30 patents. Dr. Junjie Yan is the holder of a number of distinguished titles, including: Yangtze River Scholar Professor, which is awarded by China's National Science Foundation to distinguished young scientists; the distinguished expert who is awarded the special allowance by the State Council, China; and the New Century Excellent Talent selected from Chinese universities. He is a member of the following organizations: Multiphase Flow Expert Committee for the Chinese Society of Engineering Thermophysics, Standardization Technology Committee for Steam Turbine in the Thermal Power Plant of China, Technical Committee of Technology Research Center of Chinese Thermal Power Generation, Thermal Power Generation Expert Committee of the Chinese Society of Electrical Engineering, and Automatic and Informatization Expert Committee of the Chinese Society of Electrical Engineering.

Contributors

Sankar Bhattacharya
Department of Chemical Engineering
Monash University
Clayton VIC, Australia

Xiaotao Bi
Chemical & Biological Engineering
    Department
University of British Columbia
Vancouver, British Columbia, Canada

César Huiliñir Curío
Departamento de Ingeniería Química
Universidad de Santiago de Chile
Santiago, Chile

Xiaoqu Han
State Key Laboratory of Multiphase
    Flow in Power Engineering
Xi'an Jiaotong University
Xi'an, China

Yeek-Chia Ho
Department of Civil and Environmental
    Engineering
Universiti Teknologi PETRONAS
Seri Iskandar, Malaysia

Anthony Lau
Chemical & Biological Engineering
    Department
University of British Columbia
Vancouver, British Columbia, Canada

Duu-Jong Lee
Department of Chemical Engineering
National Taiwan University
Taipei, Taiwan

C. Jim Lim
Chemical & Biological Engineering
    Department
University of British Columbia
Vancouver, British Columbia, Canada

Ming Liu
State Key Laboratory of Multiphase
    Flow in Power Engineering
Xi'an Jiaotong University
Xi'an, China

Lester Marshall
Combustion and Fuels Section
Ontario Power Generation
Ontario, Canada

Silvio Montalvo
Departamento de Ingeniería Química
Universidad de Santiago de Chile
Santiago, Chile

Shusheng Pang
Department of Chemical and Process
    Engineering
University of Canterbury
Christchurch, New Zealand

Hamid Rezaei
Chemical & Biological Engineering
    Department
University of British Columbia
Vancouver, British Columbia, Canada

Kuan-Yeow Show
Puritek Environmental Technology
    Institute
Puritek Co. Ltd.
Nanjing, China

Shahab Sokhansanj
Chemical & Biological Engineering
    Department
University of British Columbia
Vancouver, British Columbia, Canada
Agricultural and Bioresource
    Engineering
University of Saskatchewan
Saskatoon, Canada

Francisco Stegmaier
Departamento de Ingeniería Química
Universidad de Santiago de Chile
Santiago, Chile

Hua Wang
Department of Power Engineering
Henan Polytechnic University
Jiaozuo, China

Yanjie Wang
Department of Chemical and Process
    Engineering
University of Canterbury
Christchurch, New Zealand

Junjie Yan
State Key Laboratory of Multiphase
    Flow in Power Engineering
Xi'an Jiaotong University
Xi'an, China

Yuegen Yan
Puritek Environmental Technology
    Institute
Puritek Co. Ltd.
Nanjing, China

Fahimeh Yazdanpanah
Chemical & Biological Engineering
    Department
University of British Columbia
Vancouver, British Columbia, Canada

1 Recent Advances in Biomass Drying for Energy Generation and Environmental Benefits

Shusheng Pang, Yanjie Wang, and Hua Wang

CONTENTS

1.1 INTRODUCTION

Energy supply and demand data show that the world annual consumption of energy was 575 quadrillion Btu in 2015 (U.S. Energy Information Administration, 2017), of which 83% came from fossil fuels including oil, gas, and coal which released 36 billion metric tons of carbon dioxide into the atmosphere (Hausfather, 2017). The heavy reliance of human beings on fossil fuels has already caused serious consequences such as climate change due to greenhouse gas (GHG) emissions and depletion of fossil fuel resources. Biomass has been recognized as the most promising resource for future fuels and energy (Class, 1998). Biomass, in the forms of wood chips, sawdust, bagasse, grass, and agricultural residues, is generated from trees, agricultural crops or purpose-planted coppices, which absorb carbon dioxide through photosynthesis for their growth. In this way, the whole cycle – from feedstock growing through energy processing to energy consumption – is largely carbon neutral. However, biomass is of bio-origin and it commonly has initial moisture content (mc) from 50 to over 150% (dry basis) in fresh form (green). For the conversion of biomass to energy and fuels, thermochemical conversion technologies are promising technologies in the short and medium terms which include combustion, gasification (Higman and van der Burgt, 2003; Saw and Pang 2013; Sansaniwal et al., 2017), and pyrolysis (Bridgwater and Grassi 1991; Wigley et al., 2017; Wang et al., 2017).

In order to increase energy efficiency, improve energy product quality and reduce emissions in thermochemical energy conversion, the drying of biomass to the required mc is important in the development of the energy production systems. In addition, it was found that uniformity of drying also significantly affects energy efficiency in a combined heat and power (CHP) plant (Holmberg, 2007). In the development of biomass drying technologies for integration with a bioenergy plant, low-grade heat is commonly available for biomass drying. Therefore, selection of drying technologies should consider using a drying medium at low to medium temperatures (70–120°C). On the other hand, drying operations at high temperatures and with high energy efficiency may be important for a plant where high-temperature heating source is available. For assessing the biomass drying operation, overall efficiencies of energy and exergy as well as environmental impact should be considered.

This chapter will first describe the characteristics of the woody biomass, then discuss various thermochemical biomass conversion technologies, followed by the requirements on appropriate mc of the biomass for efficient and quality energy production. Finally, an assessment of biomass drying will be presented with a focus on environmental impact, energy and exergy analysis, and impact of biomass drying on bioenergy plant performance.

1.2 CHARACTERISTICS OF WOODY BIOMASS

Properties, dimensions and initial mc of the biomass vary significantly between different biomass sources (Pang and Mujumdar, 2010). For example, branches, roots, and small top ends of logs are the biomass from tree harvesting and these are normally chopped into chips with a size of 2–3 mm in thickness, 20–30 mm in width and 30–50 mm in length. In wood processing, the biomass is generated from various operation steps and its characteristics vary depending on the wood products. In sawmills, sawdust and cut-offs are generated from timber sawing and barks are from debarking. The processing of laminated veneer lumber (LVL) produces biomass in forms of small cores, waste veneers generated during round-up peeling and veneer drying, and bark during debarking.

It has been observed that the woody biomass is a loose particulate material and its size varies from 1 to 3 mm for sawdust and up to 500 mm for barks. The initial mc also varies significantly from one source to another (50–150% in harvesting and processing of wet wood). These variations need to be considered when choosing a drying technology. The size and initial mc for any type of biomass should be maintained as uniform as possible for drying and, accordingly, sizing may be needed before drying if the particle sizes of biomass vary significantly.

In addition to the variability in their physical properties, chemical compositions also vary with the biomass sources and wood species. The typical chemical compositions of woody biomass are 40–46 wt% cellulose, 23–33 wt% hemicelluloses, and 20–31 wt% lignin with a small fraction of extractives (1–7 wt%) (Walker, 2006). The chemical composition of woody biomass will affect the emissions from drying.

1.3 THERMOCHEMICAL CONVERSION TECHNOLOGIES OF BIOMASS TO ENERGY

There are a number of technologies available to convert the biomass into different energy products and these technologies can be classified into thermochemical processes (combustion, gasification, and pyrolysis) and biochemical processes (digestion, fermentation, enzyme). This review will focus on the thermochemical processes in which the biomass mc has a direct influence on the conversion efficiency and the energy product quality.

Combustion is a mature technology which converts the biomass directly to heat in the forms of hot flue gas, hot water or steam. The steam can then be used as a heating source or for power generation. A too high mc of the biomass has negative impacts on the environment due to increased GHG emissions; it also reduces conversion efficiency as a fraction of heat is consumed for water evaporation from wet biomass. In a boiler, the energy efficiency can be increased by 5–15% with increased combustion temperature by using the dried instead of wet biomass (Class, 1998). In addition, the exhaust flue gas temperature from the heat recovery system can be reduced as the dew point is lowered with the drier feedstock biomass. The optimum mc is in the range 15–25%.

In the last two decades, extensive studies have been reported on gasification (Higman and van der Burgt, 2003; Saw and Pang, 2013; Sansaniwal et al., 2017) and pyrolysis (Bridgwater and Grassi, 1991; Wigley et al., 2017; Wang et al., 2017). Gasification is a process that converts carbonaceous materials, such as biomass, into CO and H_2 based gas mixture through biomass devolatilization, reactions between the gasification agent and volatiles from biomass devolatilization as well as reactions between char and gases available. The gasification temperature is normally in the range of 700–900°C at controlled feeding rate of gasification agent (O_2, air or steam). The resulting gas mixture is called producer gas, and can be used for heat and power generation, synthesis of liquid fuels or production of pure hydrogen and chemicals. Gasification is a very efficient method for extracting energy from many different types of organic materials. The gasification technology can also be applied for conversion of "waste" materials such as municipal organic solid wastes and residuals from forestry and agricultural industries. In gasification, the biomass needs to be dried to mc ranging from 10 to 20% to increase the energy efficiency and to reduce the tar content in the producer gas (Higman and van der Burgt, 2003; Pang and Xu, 2010; Xu and Pang, 2008).

Biomass pyrolysis is a direct thermal decomposition of the biomass components in the absence of oxygen to yield an array of useful products – liquids, gases, and solids (Bridgwater and Grassi, 1991; Wigley et al., 2017; Wang et al., 2017). Fast pyrolysis with operation temperatures from 450 to 700°C and a fast heating rate has now been developed for high yields of liquid. The liquid product, which is also called pyrolysis oil or bio-oil, has the potential for biodiesel production after upgrading treatment (Bridgwater and Grassi, 1991). The challenge for biomass pyrolysis is that the bio-oil has a very complex chemical composition with high water content, which makes its purification and upgrading difficult and costly (Bridgwater and Grassi, 1991; Wigley et al., 2017; Wang et al., 2017). The mc of the biomass has a significant impact on pyrolysis in terms of conversion efficiency and quality of the bio-oil product. The moisture in the biomass will directly contribute to the water content in the bio-oil product; therefore, biomass mc in a range of 5–10% is required for pyrolysis.

In addition, wood pellet has been regarded as a useful bioenergy resource and its annual production in the world has doubled from 14.2 million tonnes in 2010 to 28 million tonnes in 2015 (Proskurina et al., 2018). Wood pellets are made of sawdust, for which mc should be 10–15% (Kim et al., 2015).

1.4   BIOMASS DRYING TECHNOLOGIES

Dryers suitable for drying of biomass include batch through-circulation dryer (perforated floor bin dryer), packed moving bed dryer (PMB), direct rotary dryer (rotary cascade dryer), indirect rotary dryer (steam-tube rotary dryer), fluidized bed dryer, and pneumatic conveying dryer (flash dryer) (Brammer and Bridgwater, 1999, Wimmerstedt, 2007). In industrial biomass drying, the following three types of dryers have been commonly used: PMB dryers (Pang and Xu, 2010; Poirier, 2007), rotary dryers (Xu and Pang, 2008, Krokida et al., 2007), and pneumatic or flash dryers (Pang, 2001; Borde and Levy, 2007; Stenström, 2017), although other types of dryers may also be used (Holmberg, 2007; Brammer and Bridgwater, 1999; Ståhl et al., 2004).

A multistage drying system can be designed to reduce emissions of volatile organic compounds (VOCs) and to increase energy efficiency. VOC emissions during drying increase with drying temperature and, in the multistage drying system, drying temperature can be lowered in all stages resulting in lower VOC emissions (Pang, 2001; Spets and Ahtila, 2004; Rupar and Sanati, 2003). A pilot scale, pressurized flash dryer has been developed by Hukkonen and his colleagues (Hulkkonen et al., 1994) for drying peat and woody biomass.

1.4.1   PACKED MOVING BED DRYER (PMB DRYER)

A sketch of the PMB dryer is shown in Figure 1.1 (Pang, 2014), in which the wet biomass is fed from the left-hand side to a moving belt which has openings allowing the drying medium (hot gas) to flow through. In order to increase the energy efficiency of the drying, the drying medium is recycled and flows downwards through the biomass bed in the second half of the dryer. In this way, the overall airflow rate is only half of that in the arrangement where the drying air always flows upwards. However, with the drying air reversal in the second half of the dryer, the exhaust gas has a lower temperature and higher humidity, thus the corresponding equilibrium moisture content (EMC) is relatively high for the bio-originated material. This indicates that if low final mc is required, the exhaust gas temperature must be kept

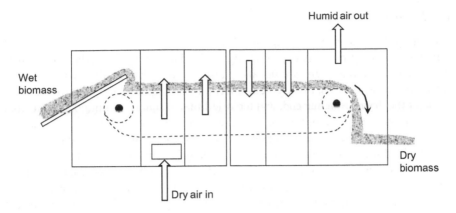

FIGURE 1.1 Schematic diagram of a packed moving bed dryer (Pang, 2014).

higher than a certain value to achieve the required dryness. However, higher exhaust air temperature results in greater heat loss.

During drying, the biomass mc varies through the bed thickness (Bengtsson, 2008), therefore, the bottom layers in the first half of the dryer dry faster than the top layers. Although the reverse flow of the drying air in the second half of the dryer reduces this variation, the mc gradient may not be totally eliminated as the recycled drying gas in the second half of the dryer has a lower drying temperature and higher humidity, thus the drying rate is lower compared to the first half of the dryer.

In the case when a low final mc is required for the dried biomass, a counter-current arrangement can be used where the hot drying gas is firstly fed from the second half of the dryer, flowing upwards, and is then reversed to flow through the biomass bed in the first half of the dryer. In this improvement, the drying efficiency is higher and the required final mc can be achieved (Pang and Xu, 2010). Based on a mathematical model developed by Pang and Xu (Pang and Xu, 2010), it has been shown that with hot air as drying medium at a velocity of 5 m/s, temperature of 120°C and absolute humidity of 0.01 kg/kg, when airflow is reversed in the second half of the dryer, the final mc varies from 18 to 21%. This is, on average, 2% higher than in the sole upwards flow but with half of the drying air mass being used. This range of mc is suitable for biomass combustion but is slightly higher for gasification and too high for pyrolysis. However, with the configuration that the drying air enters the second half of the dryer, upward flow, and is then forced to flow downwards in the first half of the dryer, the final mc of 15–18% can be achieved. This range of final mc is suitable for biomass combustion and gasification.

The advantages of the PMB dryer include its relatively simple structure, low capital cost and high drying efficiency with recycling of the drying gas. However, this type of dryer has difficulties in using a high-temperature drying medium and thus is suitable for required final biomass mc above 15%. It provides an opportunity to effectively use a low-temperature heating source (70–120°C) which is commonly available in energy processing plants.

1.4.2 ROTARY DRYER

The rotary dryer, as shown in Figure 1.2, is one of the most commonly used technologies in woody biomass drying (Pang, 2014; Pang and Mujumdar, 2010). It is effective for handling both sawdust and chips (Xu and Pang, 2008; Krokida et al., 2007; Zabaniotou, 2000). In the co-current rotary dryer, the wet material and the hot air enter the dryer from one end, and the dried material and the humid air exit from the other end. On the other hand, with the counter-current configuration, the wet material and the hot air enter the dryer from opposite ends and move inside the drum in opposite directions. In this way, the dried material can have much lower mc than that of the co-current rotary dryer because the driest solids in the counter-current rotary dryer are exposed to the gas with the highest temperature. However, in the counter-current configuration, the fire risk is much higher than it is in the co-current rotary dryer as the dry material is in direct contact with high-temperature drying air (Li et al., 2010). A commercial rotary dryer for drying wood sawdust is shown in Figure 1.2a as a schematic diagram and in Figure 1.2b as an actual dryer in use. Furthermore, a novel rotary dryer, namely roto-aerated dryer, is also available now, and has higher efficiency, greater efficacy, larger processing capacity, and shorter residence time than conventional rotary dryers (Silverio et al., 2015).

High drying temperatures of up to 500°C have been used in commercial drying of woody biomass, in which case high drying rate and high energy efficiency can be achieved. In commercial drying, the co-current arrangement can ensure that the biomass at the outlet is not over-heated to prevent fire. However, this needs reliable measurement techniques or accurate model prediction to detect the biomass temperature and mc through the dryer during drying. Several models for the rotary drying can be found in the literature (Xu and Pang, 2008; Zabaniotou, 2000; Thorne and Kelly, 1980; Cao and Langrish, 1999; 2000). Apart from temperature, the efficiency of rotary dryers can also be affected by the rotation speed of the drum and the design of flight.

Due to the high drying temperature, the exhaust air has a high temperature and high humidity; therefore, an energy recovery system can be added to increase energy efficiency.

The rotary dryer is reliable and flexible, and can dry biomass with variable sizes and initial mc. High productivity can be achieved using high drying temperatures.

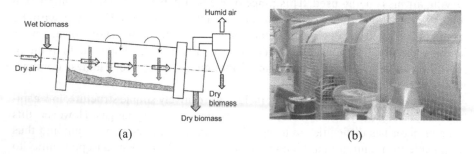

(a)   (b)

FIGURE 1.2 Rotary dryer for biomass drying: (a) schematic diagram, (b) a commercial dryer (Pang, 2014).

Moreover, the final moisture and solid temperature can be controlled through the adjustment of the flow rates of drying medium and biomass, inlet air temperature and rotation rate of the rotary drum (Shahhosseini et al., 2010). However, VOC emission is an important issue in the rotary drying of the biomass due to the high drying temperature used (Stenström, 2017). In addition, the rotary dryer is relatively expensive compared to other types of dryers.

1.4.3 Pneumatic Dryer

Pneumatic dryers are gas-solid transport systems with a convective heat and mass transfer process (Borde and Levy, 2007). This type of dryer can achieve rapid drying with short residence time by fully entraining the material in flowing gas at high velocities (Brammer and Bridgwater, 1999; Pang, 2000; 2001; Pang and Mujumdar, 2010). Figures 1.3a and b show the most common pneumatic dryer where the material residence time is from 5 to 10 s in a 100 m long pipe with single-pass for drying wood fibers or materials with small particle sizes (500 μm) using hot air or flue gas as drying medium. In the dryer, the hot gas stream transports the solid particles through a pipe or flow duct, and makes direct contact with the material to be dried. This gas stream is also a drying medium to supply the heat required for drying and carries away the evaporated moisture vapor. The low material content in the dryer enables equilibrium conditions to be reached very quickly. As the material is once-through with the drying medium, the exhaust air is humid and thus the exhaust temperature cannot be too low, otherwise condensation may occur in the cyclone for solid-gas separation. This will result in significant heat loss as well as VOC emissions from the exhaust air.

From both theoretical and experimental studies (Pang, 2000; 2001), it was found that the majority of the moisture is evaporated in the first one-third length of the dryer

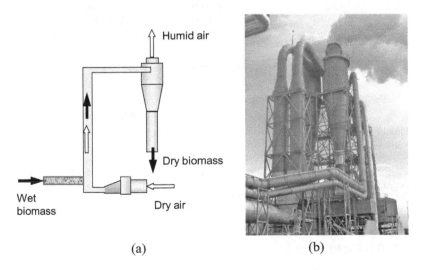

(a)   (b)

FIGURE 1.3   Pneumatic dryer for biomass: (a) a schematic diagram, (b) a commercial dryer (Pang, 2014).

and thus the dryer can be modified as a two-stage dryer to recover heat and emissions after the first stage of drying when the majority of the moisture is removed. The modified system shown in Figure 1.4 (Pang, 2000) can increase energy efficiency by up to 61% due to the lower exhaust air temperature and heat recovery from the first stage of drying (Pang, 2001). VOC emissions can also be reduced with condensate formed in the heat recovery system. Similar technology was proposed by Spets and Ahtila (2004) who found that using a multistage drying system can significantly reduce VOC emissions and increase energy efficiency in biomass drying. However, the use of three or more stages will result in much higher capital costs and complex operations.

Due to the small number of moving parts, the maintenance cost is low. The capital costs are also low in comparison with other types of dryers. However, pneumatic dryers have high installation costs and require an emission control system to reduce VOC emissions into the environment.

1.5 ENVIRONMENTAL ISSUES IN BIOMASS DRYING

1.5.1 EMISSIONS

Emission from biomass drying has been recognized as an important issue which needs to be carefully considered (Brammer and Bridgwater, 1999; Spets and Ahtila, 2004; Rupar and Sanati, 2003; Wimmerstedt, 1999). There are legislations to restrict emissions in many countries. The requirements to implement the emission reduction at a particular drying installation will depend on the location (how close to residential areas) and on local regulations. These requirements vary from country to country, but more rigorous regulations are expected to be put in place everywhere in the near future.

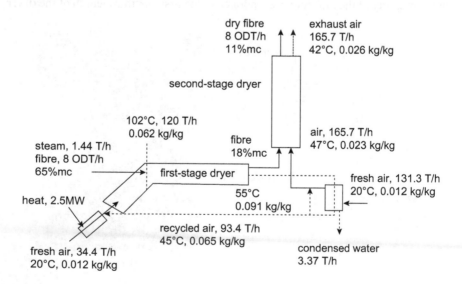

FIGURE 1.4 A two-stage wood fiber dryer with recovery of heat and VOC emission of the exhaust air from the first stage (Pang, 2000).

Wastney (1994) did a comprehensive literature review on gaseous emissions from the drying of wood and barks with the aim of characterizing the emission composition, formation, effect on the environment, and control technologies. Later reviews by other researchers (Brammer and Bridgwater, 1999; Spets and Ahtila, 2004; Rupar and Sanati, 2003; Wimmerstedt, 1999) confirmed the earlier findings by Wastney (1994).

Emissions in biomass drying come from three sources: fine particulates entrained, vaporization of volatile organic components (VOCs) in the biomass, and thermal degradation of the biomass. The vaporized components can be further categorized into those that remain volatile at ambient conditions, and those that condense at ambient conditions after drying. The most volatile components consist of monoterpenes, which are naturally emitted from wood at ambient temperatures – the emission rate increases rapidly with temperatures above $100°C$. VOC emissions can cause problems of formation of ground level ozone due to the presence of nitrogen oxide and irritation of the respiratory system due to photo-oxidants (Fagernäs et al., 2010). The condensable compounds consist of fatty acids, resin acids, diterpenes, and triterpenes (Bridgwater et al., 1995). Although these compounds have high boiling points, their vapor pressure at high drying temperatures ($180–220°C$) is the driving force for these compounds to be released from wood. Thermal degradation products, such as formic and acetic acids, alcohols, aldehydes, furfurals, and carbon dioxide, are released at higher drying temperatures ($200°C$ or higher). Increasing the wood temperature rapidly increases the degree of thermal degradation, thus promoting the release of degradation products. Some of the degradation components have a strong smell and are even suspected carcinogens if the concentration is higher than a certain level in the environment.

Emission is affected by many factors including dryer type, drying medium characteristics and biomass types. For a given dryer, the configuration of drying medium reuse or direct exhaustion to the atmosphere, the degree to which the material is agitated and broken up in the drying process, and the residence time of the material in the dryer are important factors to be considered. In addition, heat recovery and emissions abatement equipment can be installed when a low level of emission in the exhaust gas is required.

PMB dryers are designed to circulate the drying medium through biomass layers and the biomass is relatively stable on the bed, which can act as a filtration bed to trap any fines from the exhaust, resulting in low particulate emission. In addition, the residence time and temperature can be controlled to avoid any over-drying and biomass degradation. In rotary dryers, the biomass is agitated to some extent and thus significant amounts of fines are usually present in the exhaust gas. Due to the lag time of the material in the rotary dryer, over-drying may occur, resulting in higher VOC emissions. Therefore, equipment for reducing VOC emissions from the exhaust air is normally installed with the rotary dryer. In pneumatic dryers, the short residence time of biomass results in relatively low VOC emissions and degradation products. However, a highly efficient cyclone is needed to separate the dried biomass from the exhaust air.

Biomass particle size and size distribution are important for particulate emissions. A wider range of particle size distribution may result in smaller particles being over-dried, leading to excessive thermal degradation. Moreover, the storage time of the biomass is another factor that influences the amount of emissions. Milota (2013) reported that aged pine and juniper that had been stored for up to a year had significantly lower total VOC emissions than fresh ones (during drying), which was due to the volatilization of moisture and VOCs from the biomass during the storage.

1.5.2 CONTROL OF EMISSIONS IN BIOMASS DRYING

Solutions for reducing emissions from the drying of biomass have been proposed; these include improvement of dryer design, optimization of drying conditions and development of technologies to reduce emissions.

Improvements on the dryer design have been reported which are helpful for reducing emission generation. In the multistage drying system, the total emission of VOCs is lowered because of the low temperature at each stage (Silverio et al., 2015). Another common improvement is to use indirect heating where the hot steam, air or flue gas passes through the separate ducts or pipes inside the dryer. Therefore, the interactions between the drying medium and the biomass are avoided, and the impurities in the exhaust gas are significantly reduced (Li et al., 2010; Funda, 2011).

The technology of low-temperature drying is well developed recently, which is crucial for reducing emissions and drying operation costs. It is reported that industrial dryers using low-temperature heating sources are provided by some manufacturers. At the operation temperature of 30–110°C, the processing capacities of those dryers are as high as 500–4000 kg/h (Li et al., 2010). Furthermore, it is reported that it is possible to dry biomass to low mc of 10 wt% with reduced VOC emissions into the drying medium when the material temperature is maintained at 100°C or less (Fagernäs et al., 2007).

In commercial drying of biomass, various emission control technologies can be chosen depending on the form of emission to be controlled. For particulate emission control (solid), filters, baghouse filters, and multistage cyclones are commonly used. The filter is suitable for a gas flow capacity of 140–2000 m³/min. A liquid scrubber is commonly used for the treatment of gas emissions by employing a solvent. In this case, the discharged solvent also needs treatment, although an aeration pond is a straightforward solution. In most cases, there is a dilution of the solvent and discharging into the sewage system, but this is not recommended as a long term and sustainable solution. The capacity for the scrubbing treatment is similar to the filter treatment at a range of 28–1500 m³/min but the capital costs for the solvent scrubbing are higher than for the filter treatment.

Although most of the particles in the outlet gas of the biomass dryers can be removed by the above technologies, condensable volatiles are still easily released when the dryer temperature is higher than 180–220°C (Li et al., 2010). The condensable organics will form aerosol when the temperature is lowered, which is called "blue haze" (Amos, 1998). A wet electrostatic precipitator is the most effective method to remove the fine solids and aerosol in the exhaust gas (Li et al., 2010).

Both particulate and gas emissions can be cracked by regenerative thermal oxidization (RTO) by feeding the exhaust gas to a very high-temperature incinerator (above 1300°C). In this way, the volatile organic components and particulates in the gases will be cracked to lighter molecular weight compounds through the high-temperature combustion. This needs a specially designed incinerator, which can operate at such high temperatures, and further studies on a commercial scale application are needed to confirm its effectiveness.

A more attractive technology to reduce VOC emissions is to use superheated steam as the drying medium or to add a heat recovery system from the exhaust drying air with hot air drying. Using superheated steam as a drying medium has advantages such as the absence of fire or explosion risk, and has higher drying efficiency if the exhaust steam is re-utilized. In the superheated steam drying system, the steam is recycled through a heat exchanger to maintain the required temperature (Mujumdar, 2007; Pang, 1997; Pang and Dakin, 1999). The additional steam generated from moisture evaporation is condensed through a condenser and the condensate is removed from the drying system. The limitations of using superheated steam are the greater complexity of the system and its operation.

Heat recovery for the exhaust drying air is achieved by adding a heat exchanger in which the exhaust drying air is cooled down while the heat is recovered through heating boiler feeding water or preheating the fresh drying air. While the exhaust drying air is cooled down, condensate may be formed which also dissolves the VOCs. By using superheated steam drying or a heat recovery system, VOC emissions in the gas stream are significantly reduced; however, the problem is transferred to the liquid phase, although the waste liquid is relatively easy to process either by inorganic treatment or by biological treatment.

1.6 ANALYSIS OF AN INTEGRATED BIOMASS DRYING AND BIOENERGY PROCESSING PLANT

1.6.1 PLANT INTEGRATION AND ENERGY/EXERGY ANALYSIS

Recent studies have been performed to examine the exergy efficiencies and impact of biomass drying on the whole plant performance (Puladian, 2015; Puladian et al., 2016). Exergy analysis has attracted more attention recently as it reflects how efficient the quality energy utilized is. One unique feature of the biomass to energy plant is that the energy supply for drying can be from exhaust flue gas or low-temperature steam and water. In large-scale combustion boiler systems, heat and power are normally the target energy products and the power is generated using a steam turbine. The exhaust steam from the turbine is cooled down and then the condensate (water) is circulated back to the boiler. In this way, the energy extracted from the steam condensation can be used for biomass drying as shown in Figure 1.5, which is a modified version of Li et al. (2012). In this example, the exhaust air from biomass drying can be fed to the boiler as an oxidant to recover sensible energy in the hot air and to crack the drying VOCs. Therefore, there are apparent benefits of integrating biomass dryer and boiler, which include increasing boiler efficiency, reducing emissions in the exhaust gas, and improving whole plant performance (Li et al., 2012). In a boiler, energy efficiency can be increased by 5–15% with increased combustion temperature and reduced flue gas temperature by using dried instead of wet biomass (Pang, 2008).

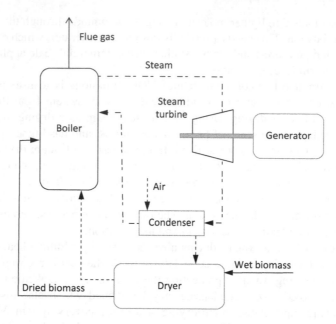

FIGURE 1.5   An integrated drying-power generation plant using a combustion boiler (Pang, 2014 modified from Li et al., 2012).

In medium- or small-scale boilers, the hot water or steam is generated for heat supply to residential heating or process plant requirements. In this case, flue gas can be introduced directly to the dryer for biomass drying. Alternatively, a heat exchanger can be installed to recover the exhaust heat of the flue gas for preheating air or water, which is then used for biomass drying (Wimmerstedt, 2007). It was reported that flue gas usage in the dryer can significantly lower the capital cost of a power plant (Li et al., 2012).

A more comprehensive integration of biomass drying with advanced biomass gasification was designed by the research team at the University of Canterbury, New Zealand, as shown in Figure 1.6. The gasifier uses steam as a gasification agent and consists of two reactors: the bubbling fluidized bed (BFB) gasification reactor and the circulating fluidized bed (CFB) combustor. The biomass is fed to the gasification reactor by a mechanical screw feeder and steam is fed from the bottom of the reactor as the gasification agent and fluidization agent. The endo-thermic gasification reactions occur with the required heat being supplied by hot, circulating bed materials which are heated in the combustion reactor where solid char from the gasification reactor is combusted. By gravity, the bed material and solid char flow from the gasification reactor to the combustion reactor through an inclined pipe, called chute. Air, as both oxidant and fluidization agent, is injected into the combustion reactor for the char combustion. The bed materials, heated to 800–900°C depending on the gasification temperature, are carried up and then out of the reactor by flue gas to a cyclone in which the flue gas and the hot bed materials are separated. The flue gas flows out from the cyclone top while the hot bed materials drop to the gasification reactor through a siphon. The main

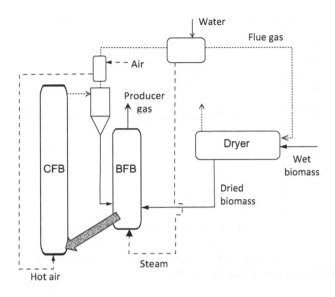

FIGURE 1.6 The process flow diagram for an integrated biomass drying and gasification system (Pang, 2014).

advantage of this dual fluidized bed (DFB) gasification system is that the producer gas has a high hydrogen content (up to 60% v/v) and thus a high calorific value (12–14 MJ/Nm³).

High energy efficiency of the system can be achieved when the clean flue gas is used for the preheating of air and generation of steam followed by biomass drying as shown in Figure 1.6. In the integrated drying-gasification design, the woody biomass enters the rotary dryer where it is dried to target moisture content of 15–20% and the dried biomass then goes to the gasifier.

1.6.2 Impact of Biomass Drying on Bioenergy Plant Performance

Mathematical models based on mass and heat balances as well as thermodynamic analysis can be developed to simulate the integrated biomass drying-energy production plant. Sensitivity analysis can then be performed using mathematical models to address key questions such as whether the available exhaust energy from the plant is sufficient for drying; what impact a fluctuation in the biomass initial moisture content would have on the plant performance; what the benefits of biomass drying are and how much additional cost is caused by adding the drying operation. An example of model simulation results is illustrated in Figure 1.7 (Puladian 2015) on the effects of biomass initial moisture content in the integrated system of a rotary biomass dryer and a DFB biomass gasifier. Figure 1.7 shows that the plant overall energy efficiency decreases dramatically with the initial moisture content of biomass whereas the exergy efficiency is reduced insignificantly. In one way, the reduction in energy efficiency is due to the increased energy loss by the exhaust gas from the dryer as higher exhaust temperature is needed to prevent condensation when humidity is increased

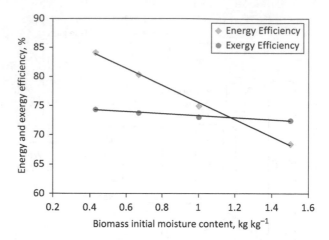

FIGURE 1.7   Effect of biomass initial moisture content before drying on plant energy efficiency and exergy efficiency (Puladian, 2015).

at higher evaporation load during drying. On the other hand, when the biomass initial moisture content is higher, more flue gas heat could be recovered for drying. Therefore, there will be less steam generated in the system and more steam should be imported, which results in the dramatic decrease of energy efficiency of the plant. However, since the exergy value of steam is much lower than its energy value, the exergy efficiency is less dependent on the amount of the imported steam (Puadian et al., 2014).

In practice, the biomass generated from forest harvesting is normally left on site for some time for natural drying. In this way, the biomass moisture content can be reduced to 50% or less. Through this method, the energy efficiency of the whole integrated system will be increased.

1.6.3   A CASE STUDY

A case study has been reported for a 100 MW biomass to liquid fuel (BTL) plant (Puladian, 2015; Puladian et al., 2016); the flow diagram for this is shown in Figure 1.8. In this plant, the biomass feedstock has an initial moisture content of 100% which is reduced to 18% during drying before gasification. In the biomass gasification, the producer gas is cleaned using rapeseed oil or other solvents for tar removal before it is fed to a liquid fuel synthesis reactor (Fischer-Tropsch process). The crude liquid is then refined for production of gasoline (petrol) and diesel. The off-gas from the Fischer-Tropsch reactor is used in a gas engine for power generation which produces 12.3 MW electricity from which 0.7 MW is exported. The target products of the plant include 19.7 MW diesel, 14 MW gasoline, and 2.2 MW fuel gas. However, the net liquid fuel yield from the plant is 28.6 MW, which is the sum of the yields of diesel and gasoline minus rapeseed oil make-up. In addition, 21 MW heat in the form of steam and 2.2 MW heat in the form of hot flue gas is generated for export from the plant.

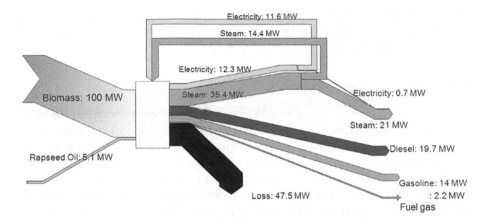

FIGURE 1.8 The energy flow diagram of the BTL plant (Puladian, 2015).

By considering all of the energy input (biomass and rapeseed oil make-up) and energy generation of the plant, the total energy efficiency was found to be 54.8%. This consists of chemical efficiency for the liquid fuel (32.1%), heat efficiency (22.1%), and electrical efficiency (0.6%). The exergy efficiency of the plant was estimated to be 39.4% which is associated with the production of the liquid fuels as target products, and heat and electricity as by-products.

The effect of the final moisture content of the biomass after drying on energy and exergy efficiencies and liquid fuel yield from the BTL plant is shown in Figure 1.9. As can be seen from the figure, with an increase in the final moisture content of the biomass from 10 to 30%, the energy efficiency increases from 53.4 to 57.2% while the net liquid fuel yield declines from 29.8 to 26.6 MW. The exergy efficiency remains relatively constant with the change in final moisture content of biomass after drying.

On one hand, with an increase in the final moisture content of the dried biomass, less heat is consumed on the drying while more steam is generated from the heat

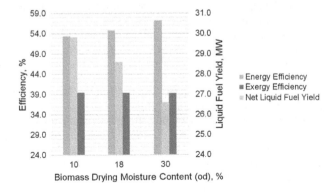

FIGURE 1.9 Impact of biomass moisture content on the bioenergy plant performance (Puladian, 2015).

recovery. On the other hand, with an increase in biomass moisture content fed to the gasifier, less steam is needed in the gasification system. Therefore, the steam export from the system is increased.

1.7 CONCLUDING REMARKS

Woody biomass, in the forms of wood chips, bark, and sawdust is one of the most promising and sustainable resources for future production of energy and liquid fuels. Conversion of woody biomass to energy products can be achieved through thermo-chemical conversion processes including combustion, gasification, and pyrolysis, which require the biomass moisture content (mc) to be within the range of 5–25%. Drying of the biomass to appropriate mc increases energy efficiency, improves energy product quality, and reduces environmental impacts in the biomass conversion. The selection of a suitable drying system and drying conditions are critical to mitigate the environmental concerns and to achieve the required final mc. In addition, with optimized drying of biomass, the overall performance of the bioenergy plant can be improved with increased energy and exergy efficiencies, and product quality.

REFERENCES

Amos, W. A. 1998. *Report on Biomass Drying Technology.* Report NREL/TP-570-25885, National Renewable Energy Lab., Golden, Colorado, USA.
Bengtsson, P. 2008. "Experimental analysis of low-temperature bed drying of wooden biomass particles." *Drying Technology* 26(5):602–610.
Borde, I., and A. Levy. 2007. "Pneumatic and flash drying." In *Handbook of Industrial Drying* (3rd Edition), edited by Arun S. Mujumdar. Philadelphia, USA: CRC/Taylor & Francis.
Brammer, J. G., and A. V. Bridgwater. 1999. "Drying technologies for an integrated gasification bio-energy plant." *Renewable and Sustainable Energy Reviews* 3(4): 243–289.
Bridgwater, A. V., and G. Grassi. 1991. *Biomass Pyrolysis Liquid: Upgrading and Utilisation.* New York, USA: Elsevier Applied Science.
Bridgwater, A. V., D. C. Elliott, L. Fagernäs, J. S. Gifford, K. L. Mackie, and A. J. Toft. 1995. "The nature and control of solid, liquid and gaseous emissions from the thermochemical processing of biomass." *Biomass and Bioenergy* 9(1–5): 325–341.
Cao, W. F., and T. A. G. Langrish. 1999. "Comparison of residence time models for cascading rotary dryers." *Drying Technology* 17(4&5): 825–836.
Cao, W. F., and T. A. G. Langrish. 2000. "The development and validation of a system model for a countercurrent cascading rotary dryer." *Drying Technology* 18(1&2): 99–115.
Class, D. L. 1998. *Biomass for Renewable Energy, Fuels, and Chemicals.* San Diego, USA: Academic Press.
Fagernäs, L., J. Brammer, C. Wilén, M. Lauer, and F. Verhoeff. 2010. "Drying of biomass for second generation synfuel production." *Biomass and Bioenergy* 34(9): 1267–1277.
Fagernäs, L., P. McKeough, and R. Impola. 2007."Behaviour and emissions of forest fuels during storage and drying." *Proceedings of 15th European Biomass Conference & Exhibition–From Research to Market Deployment.* 7–11 May 2007, Berlin, Germany.
Funda, Z. 2011. *Usage of Very Wet Biomass for Energy Production.* PhD thesis, Czech Technical University in Prague, Czech Republic.
Hausfather, Z. 2017. "Analysis: Global CO_2 emissions set to rise 2% in 2017 after three-year 'plateau'." CarbonBrief, accessed 1 Jun 2018. https://www.carbonbrief.org/analysis-global-co2-emissions-set-to-rise-2-percent-in-2017-following-three-year-plateau.

Higman, C., and M. van der Burgt. 2003. *Gasification*. Amsterdam, The Netherlands: Elsevier Science.

Holmberg, H. 2007. *Biofuel Drying as a Concept to Improve the Energy Efficiency of an Industrial CHP Plant*. PhD thesis, Helsinki University of Technology, Finland.

Hulkkonen, S., O. Heinonen, J. Tiihonen, and R. Impola. 1994. "Drying of wood biomass at high pressure steam atmosphere; experimental research and application." *Drying Technology* 12(4): 869–887.

Kim, S. H., I. Yang, and G. S. Han. 2015. "Effect of sawdust moisture content and particle size on the fuel characteristics of wood pellet fabricated with Quercus mongolica, Pinus densiflora and Larix kaempferi sawdust." *Mokchae Konghak: Journal of the Korean Wood Science and Technology* 43(6): 757–767.

Krokida, M., D. Marinos, and A. S. Mujumdar. 2007. "Rotary drying." In *Handbook of Industrial Drying* (3rd Edition), edited by Arun S. Mujumdar. Philadelphia, USA: CRC/Taylor & Francis.

Li, H., Q. Chen, X. Zhang, K. N. Finney, V. N. Sharifi, and J. Swithenbank. 2012. "Evaluation of a biomass drying process using waste heat from process industries: A case study." *Applied Thermal Engineering* 35: 71–80.

Li, H., K. N. Finney, J. Swithenbank, and V. Sharifi. 2010. *EPSRC Thermal Management of Industrial Processes: A Review of Drying Technologies*. Report by SUWIC, Sheffield University, Sheffield, UK.

Milota, M. R. 2013. "Emissions from biomass in a rotary dryer." *Forest Products Journal* 63(5–6): 155–161.

Mujumdar, A. S. 2007. "Superheated steam drying." In *Handbook of Industrial Drying* (3rd Edition), edited by Arun S. Mujumdar, 439–452. Philadelphia, USA: CRC/Taylor & Francis.

Pang, S. 1997. "Some considerations in simulation of superheated steam drying of softwood lumber." *Drying Technology* 15(2): 651–670.

Pang, S. 2000. "Mathematical modelling of MDF fiber drying: Drying optimization". *Drying Technology* 18(7): 1433–1448.

Pang, S. 2001. "Improving MDF fibre drying operation by application of a mathematical model." *Drying Technology* 19(8):1789–1805.

Pang, S. 2008. Guest editorial: "Biomass drying: Areas for future R&D needs and sustainable energy development." *Drying Technology* 26: 623–624.

Pang, S. 2014. "Biomass drying for an integrated bioenergy plant." In *Handbook of Industrial Drying* (4th Edition), edited by Arun S. Mujumdar, 847–860. Philadelphia, USA: CRC/Taylor & Francis.

Pang, S., and M. Dakin. 1999. "Drying rate and temperature profile for superheated steam vacuum drying and moist air drying of softwood lumber." *Drying Technology* 17(6): 1135–1147.

Pang, S., and A. S. Mujumdar. 2010. "Drying of woody biomass for bioenergy: Drying technologies and optimization for an integrated bioenergy plant." *Drying Technology* 28(5): 690–701.

Pang, S., and Q. Xu. 2010. "Drying of woody biomass for bioenergy using packed moving bed dryer: Mathematical modeling and optimization." *Drying Technology* 28(5): 702–709.

Poirier, D. 2007. "Conveyor dryers." In *Handbook of Industrial Drying (3rd Edition)*, edited by Arun S Mujumdar. Philadelphia, USA: CRC/Taylor & Francis, pp. 411–422.

Proskurina, S., M. Junginger, J. Heinimö, B. Tekinel, and E. Vakkilainen. 2018. "Global biomass trade for energy—Part 2: Production and trade streams of wood pellets, liquid biofuels, charcoal, industrial roundwood and emerging energy biomass." *Biofuels, Bioproducts and Biorefining*. Doi:10.1002/bbb.1958 (published online in 2018).

Puladian, N. 2015. Development of an Integrated System Model for Production of Fischer-Tropsch Liquid Fuels from Woody Biomass. PhD thesis, Department of Chemical and Process Engineering, University of Canterbury, Christchurch, New Zealand.

Puadian, N., J. Li, and S. Pang. 2014. "Analysis of operation parameters in a dual fluidized bed biomass gasifier integrated with a biomass rotary dryer: Development and application of a system model." *Energies* 7(7): 4342–4363.

Puladian, N., S. Pang, and J. Li. 2016. "Effect of biomass drying on liquid fuel yield, and efficiencies of energy and exergy in an advanced biomass to liquid fuel system." *The 20th International Drying Symposium (IDS 2016)*, Gifu, Japan.

Rupar, K., and M. Sanati. 2003. "The release of organic compounds during biomass drying depends upon the feedstock and/or altering drying heating medium." *Biomass and Bioenergy* 25(6): 615–622.

Sansaniwal, S. K., K. Pal, M. A. Rosen, and S. K. Tyagi. 2017. "Recent advances in the development of biomass gasification technology: A comprehensive review." *Renewable and Sustainable Energy Reviews* 72: 363–384.

Saw, W. L., and S. Pang. 2013. "Co-gasification of blended lignite and wood pellets in a 100kW dual fluidised bed steam gasifier: The influence of lignite ratio on producer gas composition and tar content." *Fuel* 112: 117–124.

Shahhosseini, S., M. T. Sadeghi, and H. R. Golsefatan. 2010. "Dynamic simulation of an industrial rotary dryer." *Iranian Journal of Chemical Engineering* 7(2): 68–77.

Silverio, B. C., E. B. Arruda, C. R. Duarte, and M. A. S. Barrozo. 2015. "A novel rotary dryer for drying fertilizer: comparison of performance with conventional configurations." *Powder Technology* 270: 135–140.

Spets, J. P., and P. Ahtila. 2004. "Reduction of organic emissions by using a multistage drying system for wood-based biomasses." *Drying Technology* 22(3): 541–561.

Ståhl, M., K. Granström, J. Berghel, and R. Renström. 2004. "Industrial processes for biomass drying and their effects on the quality properties of wood pellets." *Biomass and Bioenergy* 27(6): 621–628.

Stenström, S. 2017. "Drying of biofuels from the forest—A review". *Drying Technology* 35(10): 1167–1181.

Thorne, B., and J. J. Kelly. 1980. "Mathematical model for the rotary dryer." In *Drying'80*, vol. 1, 160–169. Washington DC, USA: Hemisphere Publishing.

U.S. Energy Information Administration. 2017. *International Energy Outlook 2017*. Washington, DC, USA: U.S. Energy Information Administration.

Walker, J. C. F. 2006. *Primary Wood Processing: Principles and Practice*. Dordrecht, The Netherlands: Springer Science & Business Media.

Wang, S., G. Dai, H. Yang, and Z. Luo. 2017. "Lignocellulosic biomass pyrolysis mechanism: A state-of-the-art review." *Progress in Energy and Combustion Science* 62: 33–86.

Wastney, S. C. 1994. *Emissions from Wood and Biomass Drying: a Literature Review*. Rotorua, New Zealand: New Zealand Forest Research Institute Ltd.

Wigley, T., A. C. K. Yip, and S. Pang. 2017. "A detailed product analysis of bio-oil from fast pyrolysis of demineralised and torrefied biomass." *Journal of Analytical and Applied Pyrolysis* 123: 194–203.

Wimmerstedt, R. 1999. "Recent advances in biofuel drying." *Chemical Engineering and Processing: Process Intensification* 38(4–6): 441–447.

Wimmerstedt, R. 2007. "Drying of peat and biofuels". In *Handbook of Industrial Drying* (3rd Edition), edited by Arun S. Mujumdar. Philadelphia, USA: CRC/Taylor & Francis.

Xu, Q., and S. Pang. 2008. "Mathematical modeling of rotary drying of woody biomass." *Drying Technology* 26(11): 1344–1350.

Zabaniotou, A. A. 2000. "Simulation of forestry biomass drying in a rotary dryer." *Drying Technology* 18(7): 1415–1431.

2 Biomass Drying and Sizing for Industrial Combustion Applications

Hamid Rezaei, Fahimeh Yazdanpanah,
Shahab Sokhansanj, Lester Marshall,
C. Jim Lim Anthony Lau, and Xiaotao Bi

CONTENTS

2.1 REPLACING COAL WITH BIOMASS IN POWER PLANTS

2.1.1 BACKGROUND

Conventional fossil fuels such as crude oil and coal have high energy density and stable properties that facilitate storage and transport, thus have been widely used for industrial and home applications [1]. The recent reduction in affordable fossil fuels, environmental concerns about the increased greenhouse gas (GHG) emissions from these fuels, and increased world energy demand have motivated the public and private sectors to turn to renewable energy sources and new technologies. In recent years, the North American governments adopted regulations to phase out coal-fired power plants by 2030 and investigate other resources to reduce GHG emission [2, 3].

Biomass feedstocks are considered as a renewable resource with a near-zero CO_2 input-output balance [4–6]. Biomass – recently living biological material and animal wastes – has been used since early history to cook and heat spaces where humans live and labor. Since the eighteenth century, biomass has been used to provide heat, steam, and power for work processes. Today, biomass has an expanded role in the global demand for energy. Biomass is the only renewable energy source which may be converted to all the phases of fuels: solid, liquid, and gas [7]. On the other hand, its utilization is connected to a number of technical challenges. Compared to fossil fuels, biomass is lower in energy (17–19 MJ/kg) and bulk density (60–100 kg/m^3) [8]; heterogeneous in physical, chemical, and thermal properties; high in moisture [8], mineral [9], and oxygen contents [10]; highly hygroscopic [11], and difficult to handle [8].

Bioenergy, the renewable energy produced from biomass, is a promising solution to environmental challenges and a driver of economic development from local to global levels. Bioenergy is a source of power, heat, and liquid and gaseous fuels from biomass. Converting biomass to either secondary liquid and gaseous fuels or electricity through thermal conversion processes are the ways of increasing the energy density and transportability. Uses of these various forms of bioenergy include industrial, residential, and commercial applications.

The current chapter focuses on the application of lignocellulosic biomass to replace coal in power plants where biomass is combusted to produce electricity. The compatibility of the wood pellet with existing coal combustion facilities and the required pre-processing such as pulverizing and drying of the biomass are explained using either laboratory or industrial scale case studies. Furthermore, the current chapter presents useful biomass drying analyses for research purposes.

2.1.2 BIOMASS COMBUSTION IN POWER PLANTS

A major application of lignocellulosic biomass in a thermochemical conversion process is combustion to generate electricity, steam, and heat. Wood, with fuel characteristics of low ash and low sulfur content compared to agricultural residues, allows a direct comparison with fossil fuel and provides a good heating performance at a cost that is typically lower than that of heating oil [12]. The application of woody biomass in power plants can be effected as a full capacity single fuel approach or co-firing with conventional solid fuels such as coal.

A solid biomass fuel particle introduced into a hot combustion environment dries and devolatilizes, forming a residual char. The residual char is oxidized by O_2, CO_2, and H_2O; and finally, an ash residue remains. The thermal conversion stages – drying, the release of volatiles, and char conversion – have been investigated in numerous studies. The release of volatiles and particularly char oxidation occur very differently for biomass compared to coal [13]. The volatiles yield of biomass is very high, typically 80–90%; in other words, much higher than any coal qualities. The remaining char residues are highly porous and reactive. Biomass chars are more reactive than coal chars with respect to oxidation by O_2, CO_2, and H_2O [13]. The char oxidation or "gasification" reactions by CO_2 and H_2O are important only at very high temperatures, above 1200°C, in the case of pulverized coal combustion [14]. However, for the more reactive biomass chars, char oxidation reactions by CO_2 and H_2O must be taken into consideration at temperatures as low as 800°C [15].

2.2 BIOMASS PROPERTIES IN A COMBUSTION PROCESS

The literature shows that the conversion rate and efficiency of the biomass in the combustion process strongly depend on the local temperature inside the biomass particles [16]. In an industrial combustion application, biomass particles would be exposed suddenly to the hot environment. The biomass has a broad range of properties such as moisture content, dimensions, and density which are proven to influence the efficiency of combustion. Woody biomass may be used as a combustion feedstock in the form of either wood chips or wood pellets [4, 5, 17]. Wood pellets have a higher bulk density and more homogenous physical properties than wood chips [18, 19]. The combined effects of particle properties such as size, shape, and density [20–25]; reactor temperature [26–30]; moisture content, and the heating mechanism (convection/radiation) [31], contribute to the formation of temperature profile inside the biomass particle, affecting the particles' heating rate and consequently the rate of conversion [3]. In addition to combustion rate, particle properties such as size, shape, density, and moisture content influence the handling of biomass.

The Ontario Power Generation (OPG) is an example of full replacement of coal with woody biomass in the Atikokan and Thunder Bay power generation stations. OPG published the Atikokan case study [2], which is a good illustration of the biomass process flow from storage to the combustion chamber in an existing coal power plant.

2.2.1 AN INDUSTRIAL EXAMPLE – ATIKOKAN OPG POWER GENERATION STATION

For a more efficient combustion, wood pellets are pulverized to particles smaller than 2 mm using a modified coal pulverizer. The industry is starting to agree on a value of 80–90% of particles passing the 1 mm threshold. Figure 2.1 shows the particle size distribution of pulverized pellet particles in a modified industrial coal pulverizer to handle the biomass material.

Pulverized particles flow pneumatically using the re-circulated hot gases in the pipelines feeding the combustion chamber. Figure 2.2 displays the schematic of an industrial combustion process similar to the OPG plant. The pulverized pellet particles move pneumatically using the diluted exhaust gas from the output of the

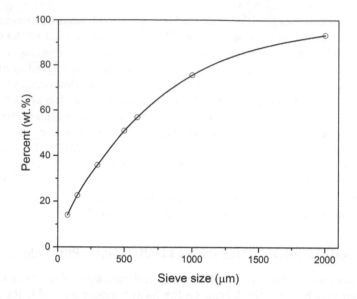

FIGURE 2.1 Particle size distribution of industrial pulverized pellet particles (pulverized by a coal pulverizer in Atikokan power station, Ontario Power Generation, OPG) [32].

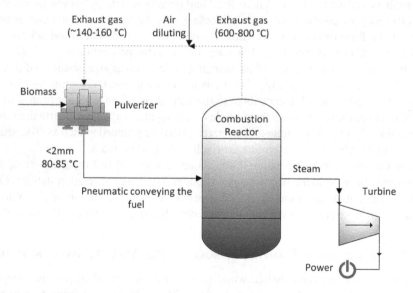

FIGURE 2.2 Schematic diagram of a combustion process, including pulverizer, pneumatic pipelines, combustion chamber, gas cleaning unit, and turbine.

combustion chamber with a temperature range of 80–85°C to prevent pre-combustion of the wood particles.

The particles exposed to hot gas flow pneumatically and dry simultaneously in the pulverizer and transport pipelines. Some of the initial moisture content of the biomass evaporates while passing through the pipelines towards the combustion chamber. The rest of the moisture is evaporated during the initial phase of devola-tilization reactions in the combustion chamber [23, 33–36]. Size distribution, shape, and density of biomass particles affect their flow properties, and kinetics of drying and thermal decomposition [18, 25, 37–39].

A necessary set of engineering data is required to design the process, such as solid handling characteristics, drying characteristics, equilibrium moisture contents, and shrinkage of particles. Sections 2.2.2 to 2.2.4 present the affecting biomass proper-ties in the flow of a combustion process such as biomass density, particle size, par-ticle shape, and moisture content. Section 4 presents the available and documented biomass drying analysis in the literature and some case studies that show the proce-dure and results of some drying research.

2.2.2 DENSITY

Biomass density influences the rate of heat and mass transfer, and consequently the conversion time of a single particle [6, 40–46]. Wood chip is produced by chipping a tree stem. The typical wood chips have a width of 30–50 mm, length of 50–100 mm, and a thickness of 5–10 mm. Pellets are produced from compacting sawdust, which typically measures 6.4 mm in diameter (using pellet die with opening diameters of 0.25 inches) and anywhere from 3 to 30 mm in length. Figure 2.3 shows a picture of wood chips and white wood pellets. Table 2.1 lists the typical dimensions and density of wood chip and wood pellet.

Wood pellet provides a form of biomass that is convenient for its handling, stor-age, and feeding to combustion boilers [17, 47]. Pellet is a promising feedstock option that is compatible with existing facilities in power generation plants [2, 48]. Although wood pellets are used mainly for oxidative combustion applications in boilers and pellet stoves at present, they may also become a potential feedstock for the pyrolysis operation to produce bio-oil.

Pelletized biomass is denser, has higher bulk density, and has more homogenous physical properties compared to the wood chip [18, 19]. Pelletization changes the properties of particles during the compression that makes the pellet a preferred

FIGURE 2.3 Picture of pine wood chip (left) and white wood pellet (right).

TABLE 2.1

Dimensions and Density of Wood Chip and Wood Pellet

Property	Wood Chip	Wood Pellet
Dimensions	Width 30–50 mm	Diameter 6.4 mm
	Length 50–100 mm	Length 3–30 mm
	Thickness 5–10 mm	
Density (g/cm³)	0.4–0.6	1.1–1.3
Bulk density (g/cm³)	0.2–0.3	0.6–0.7

feedstock for power plants compared to the wood chip. Rezaei et al. [49] show that ground pellet particles have more uniform and homogeneous physical properties than the sawdust the pellets are made of. The applied compression during pelletization changes the size and shape of particles. The ground/pulverized pellet particles have a more spherical shape – compared to the needle shape of ground chip particles – which facilitates the handling and flowability of the particles when a bulk of the material is involved. On the other hand, pelletization adds to the cost of the produced fuel per unit of mass. The raw biomass should be dried and ground to meet the required feedstock for pelletization. For the sake of comparison, the price of wood pellet may be up to twice that of the wood chips.

The literature shows that densified biomass has a lower heat diffusivity that prolongs the thermal conversion of the wood pellet. Pauilauskas et al. [50] observed a small rate of heat diffusion inside the wood pellet as well as a release of the volatiles out of the pellet compared to un-processed wood chips. The center of a pellet took about 5–10 minutes to reach the conversion temperature. Rezaei showed in two studies that both wood chip and wood pellet particles have a similar rate of drying and thermal decomposition in a slow and ramped-temperature process [32]. However, wood chip particles dried and decomposed faster than pellet particles in a fast and constant temperature process [16].

2.2.3 PARTICLE SIZE

Large pieces of biomass or biomass in the densified form (pellet and briquette) take more time to heat up and combust, which influences the conversion efficiency. This is the reason why researchers recommend to reduce the size of biomass feedstock to particles smaller than 2 mm prior to feeding a pyrolyzer or a combustion chamber (power boiler) [20–25]. A size reduction step contributes to more efficient combustion at the expense of more energy input requirement. In spite of the enormous number of studies in the literature, how to find the optimum particle size where the total consumed energy is at its minimum level is still an unanswered question.

A serious challenge for replacing coal with renewable biomass fuel in power generation stations is to modify the existing facilities to be compatible with the new fuel. One of the most important facilities to be modified is the coal pulverizer. The industrial coal pulverizer is a roller mill that crushes the fuel based on the fracture mechanism. Ontario Power Generation (OPG) reported the challenges and the implemented modifications in Atikokan power station to replace coal with white wood pellet [2]. The following industrial case study briefly explains the differences between coal pulverizing and wood pellet pulverizing, and the reason why wood pellet is the only renewable fuel compatible with the existing coal pulverizer.

2.2.3.1   An Industrial Case – Biomass Pulverizing in Existing Coal Pulverizer

The engineers in Ontario Power Generation (OPG) reported their experience in modifying the existing coal pulverizer located in Atikokan power station to be compatible with wood pellet fuel on a commercial basis [2]. The OPG team reported that three critical issues must be considered in the case of using roll-race or ball-race pulverizers to crush wood pellets instead of conventional coal fuel.

2.2.3.1.1   Limited Size Reduction

Coal pulverizers depend on fracture mechanics to grind coals to particle sizes in the 75-micron regime. However, the fibrous nature of woody biomass in the form of chips does not lend itself to this mechanism. That is the reason the coal pulverizer cannot crush a wood chip with a highly fibrous structure. The grinding elements in a traditional coal mill are expected to reduce the size of wood pellet back into its constituent dust. Although the dust used to form the biomass pellets in the range of particles smaller than 2 mm seems to be a suitable particle size distribution for stable pneumatic transport and efficient combustion, the limited size reduction of white pellets still poses a significant problem for pneumatic transport and combustion. Regarding the illustrations that OPG provided, the wood briquette with a larger internal particle size distribution may not be an appropriate feedstock to be crushed by a coal pulverizer.

2.2.3.1.2   Higher Primary Air Requirements

The saltation velocity is the minimum velocity to avoid dropout of the particles in the lines. As explained earlier, coal pulverization produces particle sizes in the 75-micron regime, but pellet pulverization crushes the fuel to its constituent dust with a size range of <2 mm. Although the density of sawdust may be lower than that of coal, the pulverized wood particles are much larger and heavier than pulverized coal and consequently need a larger lift force for a stable pneumatic transport in the mill and burner lines. As an example, in the initial runs in Atikokan power station, the larger pulverized wood pellet particles – compared to pulverized coal – tend to accumulate within the mill during the normal operation as there is insufficient lift velocity available within the mill body to expeditiously remove these large particles. This represents a potential safety concern over the long term if friction within this large re-circulating bed were to generate enough heat to pose a fire hazard.

2.2.3.1.3   Cold Primary Air

In order to recover the energy of the exhaust gas from a boiler that has a temperature of 600–800°C, this gas mixing with a stream of cooler air is used to pneumatically transport the fuel into the mill and pipelines. In the case of coal transport, the temperature of re-circulated gas was reduced to 280–300°C. However, woody biomass starts to release volatile matter at temperatures as low as 150–180°C [51, 52]. Some installations have addressed this safety issue by employing re-circulated flue gas to lower the oxygen content of the transport medium within the mills. At Atikokan, it was decided to use cold primary air to avoid the issue of early volatile matter release. Therefore, the mill inlet temperature is about 140–160°C and the mill outlet temperature is held in the 80–85°C regime for dedicated milling of wood pellets and to prevent pre-combustion of wood particles. Although the temperature selection is more about pre-heating the fuel and thermal energy at the burner tip, the drying of biomass fuel also happens, to a certain extent, and helps obtain a more efficient burning process.

2.2.4 MOISTURE CONTENT

As explained earlier, biomass particles are dried for a more efficient combustion in power plants [4, 6, 53]. A fresh biomass has a high moisture content of up to 80% [54]. High moisture content reduces the heating value of the fuel. There is a direct and strong relationship between how dry biomass fuel is and its energy content, or calorific value. Higher moisture content shifts the ignition point to higher temperatures [8, 55] because it inhibits the rise of temperature inside the particles [26, 56]. Orang and Tran [55] reported that biomass samples containing 40% moisture can ignite readily at 800°C, but they take a much longer time to ignite and to burn at 500°C and do not ignite at 400°C; whereas, the 30% moisture content in biomass readily burns at 500°C.

The optimum biomass moisture content for a combustion process depends on the balance of the combustion conversion rate and the cost of the feedstock drying. Literature states that biomass should be dried down to a moisture around 30% wet basis [55, 57]. Fuel moisture higher than around 30% reduces the combustor temperature. Reduction in combustor temperature may result in an incomplete combustion of the fuel giving rise to the emission of tars and creosote which may condense in the flue, especially if it is long or includes changes of direction, and particulates. The water may also re-condense in the flue, and all these factors may lead to corrosion of the flue and gradual accretion of material, creating a potential for eventual blockages or fire. On the other hand, although all the moisture must be driven off prior to the first stage of the combustion process, over-drying of the wood fuel below 30% by a separate drying operation adds to the cost of the operation and reduces thermal efficiency. Pre-heating and drying wood fuel using recovered energy from exhaust gas improves the thermal efficiency of the process and boosts the thermal energy at the burner tip.

Although an important advantage of the wood pellet is its low moisture content, wood pellets may be re-wetted due to poor storage and being transported or stored in a very humid environment which contributes to moisture adsorption. Open literature shows that particle size [58], density [51], drying temperature [41, 58, 59], and relative humidity of drying gas [60, 61] affect the drying rate. In Section 2.3, a comprehensive analysis of biomass drying and some case studies are explained.

2.3 BIOMASS DRYING ANALYSIS

2.3.1 FUNDAMENTAL MECHANISMS

The moisture in a solid may be either unbound or bound. Unbound moisture is removed through evaporation and vaporization. Evaporation is the process by which a liquid phase changes to vapor at a temperature below the boiling point; thus, it is a slow process. Evaporation only takes place on the surface and depends on surface area, airflow speed, relative humidity, and temperature. Evaporation occurs when the vapor pressure of the moisture on the solid surface is greater than the vapor partial pressure in the main stream of the drying medium. Vaporization is the process by which a liquid phase changes to vapor at its boiling point. For pure water, vaporization occurs at a constant boiling point temperature. For a moist biomass particle, vaporization occurs at wet-bulb temperature or higher. As vaporization happens through the bulk, not only the surface, the rate of vaporization does not depend on the surface area, wind speed, or relative humidity. In this case, the vapor pressure of the moisture over the solid is less than the atmospheric pressure [62]. In vaporization, drying is carried out by passing warm air over the product. Along the gas flow direction, the drying gas temperature is decreased and gas humidity is increased (Figure 2.4).

Handbook of Industrial Drying [62] explains the mechanisms of drying of a moist solid. From a theoretical point of view, a drying process consists of two distinguished phases: a constant rate period and a falling rate period. Figure 2.5 shows the stages of a drying process happening in a single biomass particle. During the constant rate period, free surface water vaporizes. A constant rate of moisture diffusion in air-moisture interface controls the rate of drying. Some shrinkage might occur at this stage, but it is mostly negligible for small particles (stage 1 in Figure 2.5). Towards the end of the constant rate period, moisture moves from the inside of the solid to the surface by capillary forces. The migration of water to the surface still keeps the

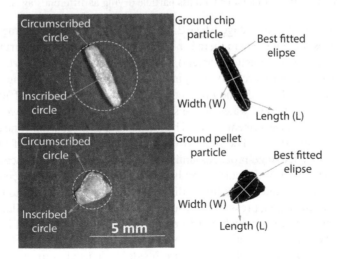

FIGURE 2.4 Image analysis of ground chip and ground pellet particles [49].

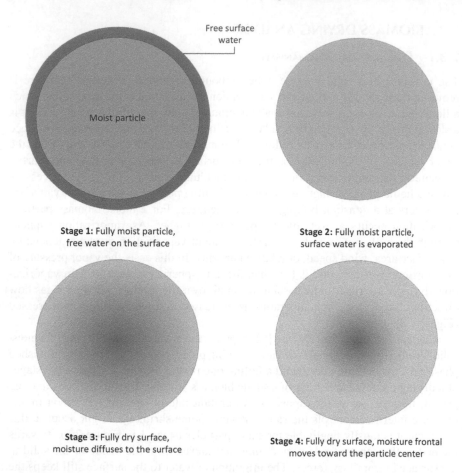

FIGURE 2.5   Schematic of a moist biomass particle drying at different stages.

drying rate constant at this stage (stage 2 in Figure 2.5). When the average moisture content reaches the point where there is not enough free water on the surface to keep a constant rate (critical moisture content), the drying rate drops and starts the falling rate period (stage 3 in Figure 2.5). In a pure diffusion process, the falling rate is supposed to be linear. However, the involved mechanisms of moisture diffusion to the surface and evaporation of water on the surface as well as within the wood particle are too complex to have a linear curve. The falling rate is mostly non-linear and may even have two sub-stages. The first falling rate stage is the period of unsaturated surface drying. This stage proceeds until the liquid film on the surface is entirely evaporated. This part of the curve may be missing entirely, or it may constitute the whole falling rate period. In the first falling rate, moisture movement to the surface happens due to diffusion, pressure build-up due to shrinkage, and capillary flow. On further drying, the rate at which moisture may move through the solid is due to the diffusion under moisture gradient and therefore pure diffusion controls the drying rate. Since the frontal part of the moisture inside the particle moves progressively towards the center of the particle and the heat conductivity of the dry external zones is smaller than in the moist zone, the drying rate increasingly drops through the heat conduction. Moisture diffusion from the inside of the particle to the surface followed by mass transfer from the surface controls the drying rate. During this stage, some of the moisture bound by sorption is removed. As the moisture concentration is lowered by drying, the rate of internal movement of moisture decreases. The rate of drying falls even more rapidly than before and continues until the moisture content falls down to the equilibrium value for the prevailing air humidity, and then drying stops.

Rezaei [3] conducted a comprehensive study on drying of woody biomass in the moisture content range of 0.1–0.9 kg water/kg dry mass basis and showed that the drying rate of wood particles had four distinct periods: a rising rate period, a short constant rate period, and two falling rate periods. Figure 2.6 illustrates the "drying characteristic curve" showing the effect of initial moisture content on the drying rate of ground wood chip particles at a temperature of 100°C. During the initial rising rate period, the particles were warmed up and the drying rate increased to a maximum value. Higher moisture content, larger particles, and lower drying temperature prolonged the warming time. Part of the free water evaporated in the warm-up period and resulted in a short constant rate period. The remaining part of the surface moisture evaporated in the constant rate period. In the falling rate period, which started with a moisture content of 0.50 (kg water/kg dry mass basis), the drying rate dropped down and approached the equilibrium moisture content. Unlike some earlier experiments reported on drying [58, 60], the switch of falling rate stage 1 to falling rate stage 2 was not very clear and there was a slight shifting. Migration of internal moisture to the particle surface during the first falling rate stage occurred through moisture diffusion, capillary flow, and internal pressure set up by an increased temperature during drying [62]. In the second falling rate stage,

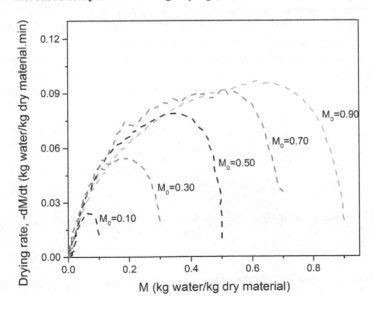

FIGURE 2.6 Effect of initial moisture content on the drying rate of ground chip particles (Tf = 100°C; carrying gas = atmospheric air) [3].

the capillary flow of moisture ceased and internal moisture diffusion controlled the rate of drying [58, 60, 62].

Handbook of Industrial Drying [62] defines the critical moisture content (Mcr) as the moisture content that the constant rate period transitions to the falling rate period. Critical moisture ratio (Mcr/M0) at the drying temperatures of 50–150°C was in the range of 0.65–0.75. The acquired values agree with the numbers in the literature [40, 41, 45, 58]. However, the moisture content of 0.1 kg water/kg dry material was way below the fiber saturation point, no constant rate period was observed, and critical moisture content may not be very practical for such a moisture content. The critical moisture ratio at 200°C drying was mostly in the range of 0.50–0.60. This could be explained by the slight degradation of biomass particles at 200°C which lost their dry mass. Figure 2.7 shows small cracks that appeared in the structure of particles dried at 200°C. At 200°C, smaller particles lost their moisture earlier and also lost parts of their dry mass.

2.3.2 Drying Rate Equations

As explained earlier, a drying process may have up to three distinct parts of a warm-up period, a constant rate period, and one or more falling rate period(s). Based on the mechanism of drying, the appropriate model should be selected to predict the drying rate.

In the warm-up period, the biomass temperature increases with time. Therefore, an unsteady-state heat transfer model should be coupled with a temperature-dependent equation for the rate of moisture evaporation from the surface. The warm-up period is usually short and would be ignored for the sake of simplicity. In the constant rate period, biomass is still beyond its fiber saturation point and its temperature is at wet-bulb temperature; thus, a constant rate of surface evaporation continues until the biomass reaches the critical moisture content.

The falling rate period is the major part of the drying process and published literature mostly focuses on this part. As the drying rate drops continuously during the falling rate, the kinetic analysis should be conducted for this part. The

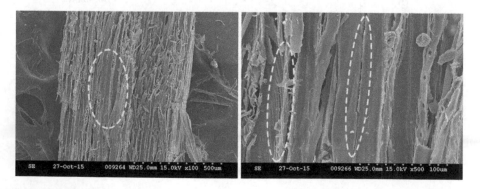

FIGURE 2.7 Scanning electron microscopic (SEM) picture of chip particles dried at 200°C (red dashed circles show the cracks in the structure of the particle) [52].

widely used kinetic models for the falling rate period are based on a moisture diffusion theory or a first order exponential model (lumped approach). Neither of the kinetic models for the falling rate period involve heat transfer; therefore, they cannot explain the high drying rate of the moisture contents beyond the critical moisture content.

For a better explanation, both approaches to modeling the falling rate period which are followed in the literature are explained in the following sections.

2.3.3 MOISTURE KINETIC ANALYSIS – DIFFUSION MODEL

The first method is the analytical solution of Fick's law. Fick's second law showing in Equation (2.1) is an unsteady-state diffusion mass transfer model.

$$\partial M / \partial t = D.\left(\partial^2 M / \partial t^2\right) \tag{2.1}$$

where

t (s)	is the process time,
x (m)	is the distance,
D (m²/s)	is the effective moisture diffusivity inside the particle, and
M (kg water/kg dry material)	is the instantaneous dry basis moisture content.

Fick's equation is mostly solved in spherical coordinates and an average particle size is used as a representative and equivalent particle diameter. As explained earlier, Fick's law is applicable for biomass with a moisture content below the critical moisture content at which the falling rate occurs. It is important to mention that the calculated moisture diffusion coefficient is an "effective moisture diffusion coefficient" which combines all phenomena explained in Section 2.3.1 for the falling rate period.

The moisture loss data in the falling rate period should be fitted to the analytical solution of Fick's second law to calculate the moisture diffusion coefficient. Fick's model is solved analytically in the spherical coordinate. Based on the assumption that the moisture diffusion coefficient does not depend on the moisture content, Equation (2.2) is the analytical solution of the diffusion equation in the spherical coordinate.

$$MR = \left(6 / \pi^2\right) \sum \left(1 / n^2\right).\exp\left(-t.\pi^2 D / R_p^2\right) \tag{2.2}$$

where

R_p(m)   is the equivalent radius of a spherical particle.

By truncating the right side of Equation (2.2) to its first term, it reduces to Equation (2.3).

$$MR = \left(6 / \pi^2\right)\exp\left(-t.\pi^2 D / R_p^2\right) \tag{2.3}$$

2.3.4 Moisture Kinetic Analysis – First Order Exponential Model

The second model is the first order mass loss equation. Similar to the diffusion model, this equation does not involve heat transfer, and drying rate constant (k) represents all phenomena explained in Section 4.1 for the falling rate period.

$$dM / dt = k(M - M_e) \qquad (2.4)$$

where
   M_e (kg water/kg dry material) is the dry basis equilibrium moisture content.

After integrating Equation (2.4), one obtains Equation (2.5).

$$M = M_e + (M_0 - M_e).\exp(-k.t) \qquad (2.5)$$

where
   M_0 kg water/kg dry material) is the dry basis initial moisture content.

Rezaei [3] showed the procedures of using both first order and diffusion models. The mass loss data was smoothed and converted to the corresponding moisture ratio (MR) using Equation (2.6).

$$MR = \frac{(M - M_e)}{(M_0 - M_e)} \qquad (2.6)$$

Drying rates ($-dM/dt$) were calculated using the finite difference method. Moisture ratio data versus time were fitted with four semi-theoretical mathematical drying models listed in Table 2.2 [40]. The Page, Henderson, and Logarithmic equations are modified versions of Newton's first order model.

The temperature dependency of the drying rate constant was formulated using Arrhenius relation.

TABLE 2.2
Semi-Theoretical Drying Kinetic Models for
Thin-Layer Isothermal Drying Condition [51]

Model	Isothermal Drying Model	Parameters
Newton	$MR = \exp(-k.t)$	k
Page	$MR = \exp(-k.t^n)$	n, k
Henderson	$MR = b.\exp(-k.t)$	b, k
Logarithmic	$MR = a + b.\exp(-k.t)$	a, b, k

$$k = k_0 . \exp(-E / RT) \qquad\qquad (2.7)$$

where

$k_0 (1/\min)$ is the pre-exponential factor,

$E(J/\mathrm{mol})$ is the activation energy,

$T(K)$ is the drying medium temperature, and

$R(8.314 \text{ J/mol } K)$ is the universal ideal gas constant.

The linear form of drying rate expressed by Equation (2.4) shows that the first order equation is suitable for a linear falling rate period. Although the falling rate period does not follow a linear form in the drying characteristic curve, the Newton model, Equation (2.5), is widely used for industrial process modeling. The reasons are that the Newton model is simple (the single parameter and the unit of k parameter are consistent).

2.3.5 A Case Study – Effects of Biomass Properties on Drying Rate

Experiments were conducted by the Biomass and Bioenergy Research Group (BBRG) to investigate the drying of ground wood chips and ground wood pellet particles at various grinding levels (grinder screen sizes of 3.2, 6.3, 12.7, and 25.4 mm), different initial moisture contents (0.1–0.5 kg water/kg dry mass basis corresponds to 9–33% wet mass basis) and drying temperatures of 50–200°C [51]. The samples were re-wetted to the specified moisture contents. For further details, please refer to Rezaei et al. [51].

It is important to mention why the drying of the wood pellet as a dry product (below 10% moisture) is also studied. It has been explained in Section 2.2.4.4 that wood pellets may be re-wetted as a result of poor storage and being transported or stored in a very humid environment, which may increase the moisture content of pellets up to 20–25% wet mass basis. Although such a range of moisture content may be acceptable for the combustion process and a separate drying operation is not necessary, the pneumatic transportation of wood fuel to the combustion chamber using hot air for boosting the thermal efficiency dries the material, too. Please refer to Section 2.2. Therefore, an experimental and numerical knowledge on the drying of both sawdust and pulverized pellet particles is necessary.

The experiments were conducted in a specially designed macro thermogravimetric analyzer (macro TGA). Ground wood chip and ground wood pellet particles are two of the available feedstocks for power generation plants. Figure 2.8 shows the schematic of the constructed macro TGA apparatus. The equipment consists of three sets of components: dryer/pyrolyzer section, instantaneous mass-measuring unit (Sartorius balance, Quintix 412–1S), and heating/controls system. The heating system includes an in-line pre-heater and an infrared heater. Rezaei [3] described the details of the designed macro TGA dryer and its operation.

Figure 2.9 shows the particle size distribution (PSD) of the ground chip and ground pellet particles. PSD of wood chip particles varied significantly with grinder screen size, but PSD of wood pellet particles was already independent of grinder

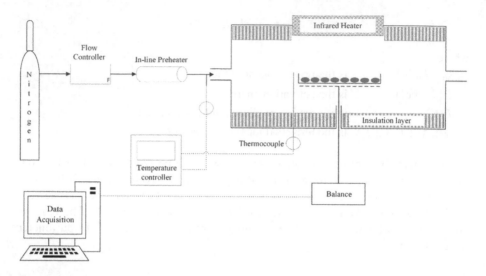

FIGURE 2.8 Schematic of the thin layer dryer (different parts are not on the same scale) [51].

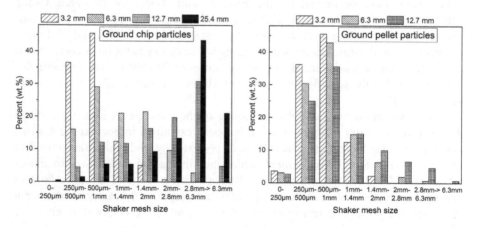

FIGURE 2.9 Particle size distribution of ground chip particles and ground pellet particles using grinder screen sizes of 3.2, 6.3, 12.7, and 25.4 mm [49].

screen size. The rate of drying decreased with an increase in the size of ground particles (Figure 2.10). Drying had a rising rate period following a falling rate period.

The experimental drying data (Figure 2.11) showed that pellet particles have a lower drying rate (Figure 2.11). Pellet particles went under a compression force during pelletization and had the lower void fractions in their internal structure. Pycnometer density of pellet particles was higher than for chip particles (Figure 2.12).

The fitting analysis showed that the Page and then Newton models were the best fitted models for the drying experimental data (Figure 2.13). In the Page model, the power of "n" varies randomly with the operating conditions (in the range of 1–2). As the value of "n" identifies the unit of drying constant (k), the Page model is not recommended for use in a comparative study. That is the reason why the Newton model

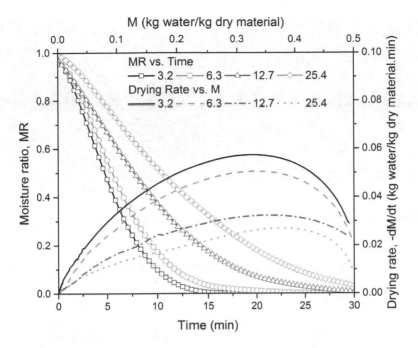

FIGURE 2.10 Moisture loss and drying rate of chip particles (grinder screen sizes of 3.2, 6.3, 12.7, and 25.4 mm; $M_0 = 0.50$ kg water/kg dry material; drying temperature of 100°C) [51].

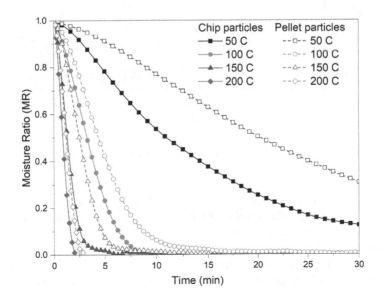

FIGURE 2.11 Moisture loss of ground chip and ground pellet particles ($M_0 = 0.30$ kg water/kg dry material; T = 50, 100, 150, and 200°C; carrying gas = atmospheric air) [52].

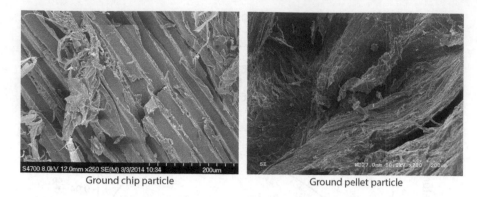

Ground chip particle   Ground pellet particle

FIGURE 2.12   SEM images of a single ground pine chip and a single ground pine pellet particle [3].

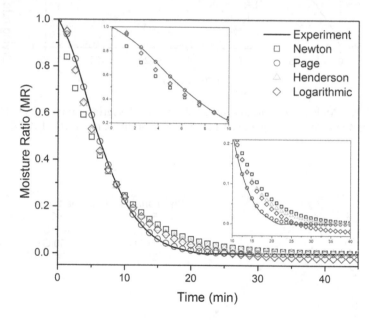

FIGURE 2.13   Fitting of four drying models for the experimental drying data of sawdust ($M_0 = 0.50$, T = 100°C) [3].

is usually employed for the sake of consistency. However, drying of pine particles exhibited an initial warming up period followed by a short constant rate period that deviated from the first order Newton model.

Rezaei [3] developed the empirical correlations to express the drying rate constant (k) as a function of drying temperature, biomass initial moisture content, and grinder screen size. As the Newton model was used, the unit of k is consistent as min^{-1}.

$$k_{\text{chips}} = \exp\left[(0.013T) - (2.372M_0) - (0.035d_{\text{gs}}) - 2.095\right] \qquad (2.8)$$

$$k_{\text{pellet}} = \exp\left[\left(0.012T\right) - \left(1.278M_0\right) - \left(0.034d_{\text{gs}}\right) - 2.533\right] \tag{2.9}$$

where

$T\left(°C\right)$ is drying temperature,

M_0 (kg water/kg dry material) is biomass initial moisture content, and

d_{gs} (mm) is grinder screen size.

Figure 2.14 shows the deviation of the proposed empirical models from experimental drying rate constants. The proposed correlations did not include the effect of carrying gas humidity. This may be one of the correlation's limitations.

The analysis of variance (ANOVA) was conducted on the experimentally obtained drying rate constants (Table 2.3). The results reveal that although the effect of the drying temperature was greater than the other two factors (M_0 and d_{gs}), the effects of all three parameters are significant. The value of P in ANOVA table represents the level of significance of the studied parameter. If the P-value of a parameter is below 0.05, it means the effect of that specific parameter is significant on the output. All three P-values are much less than 0.05 that fall in the significant zone.

To evaluate the effect of the particles' structure on the drying rate of particles, the moisture diffusion coefficient was determined using the linear form of Fick's second law for both chip and pellet particles, a linearized form of Equation (2.3). The results of the numerical analysis in Figure 2.15 show that the structure of the pellet particles was more compact and denser than in chip particles, which reduced the moisture diffusivity inside the pellet particles. Pellet particles with lower heat and moisture diffusivity took longer to dry than chip particles.

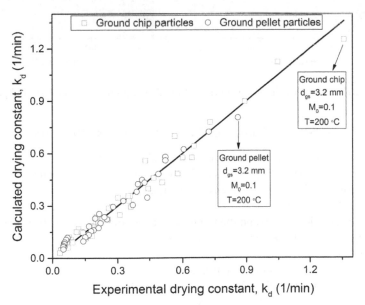

FIGURE 2.14 Agreement of experimental drying constants of the ground chip and ground pellet particles with proposed empirical models [51].

TABLE 2.3
ANOVA Table Showing the Significance of the Operating Parameters on Drying Rate Constant (k) of Chip Particles [3]

Source	DF[a]	Adj. SS	Adj. MS	F-value	P-value
Drying temperature (T, °C)	2	2.2437	0.74791	47.74	0.000
Initial moisture content (M_0, kg water/kg dry material)	2	0.5920	0.29601	18.89	0.000
Grinder screen size (d_{gs}, mm)	3	0.3415	0.11385	7.27	0.001
Error	28	0.6110	0.01567		
Total	35	3.7883			

[a] DF, degree of freedom.

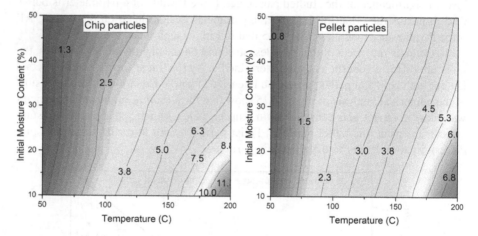

FIGURE 2.15 Moisture diffusion coefficient inside the ground chip and ground pellet particles [3].

2.3.6 A Case Study – Drying of Wood versus Bark Particles

This case study investigates the drying rate and duration of the ground wood chip (GC) and ground bark (GB) particles [63]. The rate of drying changed with the initial moisture content (0.3–0.9 kg water/kg dry mass basis corresponds to 9–47% wet mass basis) and drying temperature (70–160°C). Figures 2.16 and 2.17 show the particle size distribution (PSD) and the microscopic pictures of ground bark and ground wood chip particles. Bark particles have a broader PSD and more spherical particles than ground chip particles.

The experimental results (Figure 2.18) show that bark particles dry more slowly than wood particles. By increasing the drying temperature from 70 to 160°C, the drying time of GW decreases from 17.0 to 5.0 minutes, while the drying time of GB particles decreases from 22.0 to 6.5 minutes. The pycnometer density measurement shows that GB particles have a lower void fraction in their

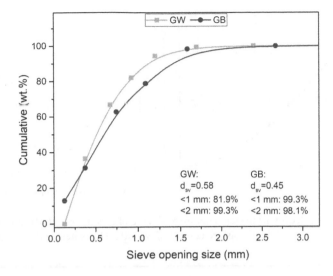

FIGURE 2.16   Particle size distribution of ground bark (GB) and ground wood (GW) particles [63].

internal structure. The denser structure of GB particles prolongs their drying process compared to the GW particles. After calculating the drying rate constant (k), the kinetic parameters (k_0 and E) are determined by the linearized form of the Arrhenius equation. Table 2.4 lists the drying kinetic parameters of both bark and wood particles with an initial moisture content of 0.9 kg water/kg dry material. Bark and wood particles have a similar activation energy that shows both have a similar sensitivity to the drying temperature. Nevertheless, the pre-exponential factor (k_0) of wood particles is significantly higher than the k_0 of GB particles that elevates the rate of drying.

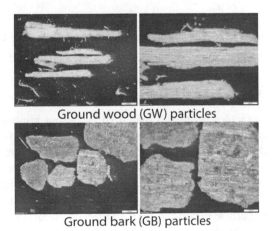

FIGURE 2.17   Microscopic picture of ground bark (GB) and ground wood (GW) particles [63].

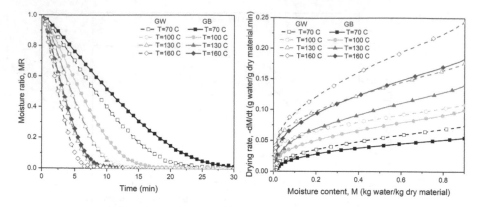

FIGURE 2.18 Effect of drying temperature on drying of ground bark (GB) and ground wood (GW) particles with an initial moisture content of $M_0 = 0.90$ kg water/kg dry material [63].

TABLE 2.4

Drying Kinetic Parameters of Ground Bark (GB) and Ground Wood (GW) Particles ($M_0 = 0.90$) [63]

Material	Drying Temperature (°C)	Drying Rate Constant, k (1/min)	Activation Energy, E (kJ/mol)	Pre-exponential Factor, k_0 (1/min)	Drying Time (min)[a]
Ground bark particles	70	0.0796	16.479	25.005	22.0
(GB)	100	0.1386			12.0
	130	0.1929			8.8
	160	0.2646			6.5
Ground wood particles	70	0.1047	16.654	38.064	17.0
(GW)	100	0.1900			9.0
	130	0.2808			6.3
	160	0.3484			5.0

[a] Drying time is the time in which material loses 90% of its initial moisture (MR drops from 1.0 to 0.1).

2.3.7 BIOMASS SHRINKAGE IN DRYING

Wood is a hygroscopic porous media that shrinks during removal of the internal bound moisture below the so-called fiber saturation point (around 0.3–0.4 kg water/kg dry material). Dehydration of fresh wood causes more reduction in the dimension of wood in a direction normal to the microfibril orientation, whereas the longitudinal shrinkage is usually up to two orders less, thus it is negligible [62, 64]. Mazzanti et al. [65] showed that the longitudinal shrinkage for poplar wood is negligible.

Shrinkage in the particle dimension itself influences the drying rate. Shrinkage is usually expressed as a shrinkage coefficient, change in dimension divided by the initial dimension [62]. Dimensional variation (shrinkage/swelling) of wood is one of the most important physical/mechanical properties that need to be better understood in order to improve its utilization in thermal or biochemical reactors [64]. Shrinkage is influenced by the particles' initial moisture content (if t the initial moisture content is equal or less than fiber saturation point) and temperature program [66], which changes the particle's diameter, specific heat, and conductivity [67]. Kowalski et al. [68] showed that drying modeling of shrinkable material needs some adjustment and pure exponential functions do not completely describe the process. Thakor et al. [69] modeled the drying rate using a first order moisture loss model in two cases of fixed radius and variable radius. They showed that the variable radius model fitted the experimental data better than the fixed radius model.

The open literature shows that rather than initial moisture content, level and direction of shrinkage depend on the particle structure and orientation of fibers. The microstructure analysis of ground chip and pellet particles showed that the ground chip particles are long in the direction of fiber length while ground pellet particles are randomly oriented [49]. Thus, it is expected that different shrinkage behavior between chip and pellet particles should be observed.

2.3.8 General Shrinkage Calculations

Volumetric shrinkage of a single particle is defined as the variation of its volume over its initial volume [70]. Using conventional methods, the dimensions of the biomass pieces are measured before and after drying to calculate the material volumetric shrinkage. In a more advanced approach, the dimension of the biomass piece should be measured during the drying process to correlate the shrinkage as a function of moisture content. Shrinkage in bulk volume and particles is defined as the reduction in volume over the initial volume, using Equations (2.10) and (2.11), respectively.

$$S_b = \frac{\left(V_{b,\text{moist}} - V_{b,\text{dry}}\right)}{V_{b,\text{moist}}} \qquad (2.10)$$

$$S_p = \frac{\left(V_{p,\text{moist}} - V_{p,\text{dry}}\right)}{V_{p,\text{moist}}} \qquad (2.11)$$

where

S_b (dimensionless)   is shrinkage coefficient of bulk volume,
S_p (dimensionless)   is shrinkage coefficient of particle volume,
$V_{b,\text{moist}}$ (cm^3)   is bulk volume of moist particles,
$V_{b,\text{dry}}$ (cm^3)   is bulk volume of dry particles,
$V_{p,\text{moist}}$ (cm^3)   is particle volume of moist particles, and
$V_{p,\text{dry}}$ (cm^3)   is particle volume of dry particles.

2.3.9 Shrinkage Calculations for Small Pulverized Biomass Particles Produced for Combustion Application

For large pieces of biomass, dimension measurement techniques such as ruler and caliper are available to measure the dimensions and consequently volume of pieces. The challenge is the shrinkage calculation of very small particles produced for combustion purposes. The reasons are that the particles are small and have a broad size distribution, and that monitoring the dimensions of individual particles is barely possible. Rezaei et al. [52] developed a novel technique to indirectly calculate the shrinkage of ground biomass particles using the measurement of average particle size of a bulk of particles. If particles are assumed in the spherical coordinate, the volume of spheres is replaced with $V_p = (\pi/6)d_p^3$. Therefore, Equation (2.11) reduces to Equation (2.12).

$$S_p = \frac{\left(d_{p,\text{moist}}^3 - d_{p,\text{dry}}^3\right)}{d_{p,\text{moist}}^3} \tag{2.12}$$

where

d_p moist (mm) is average particle diameter of ground biomass particles in moist condition, and

d_p dry (mm) is average particle diameter of ground biomass particles in moist condition.

The size distribution and average particle size for a bulk of ground particles may be measured using various size analysis methods such as sieving or laser diffraction size analyzer. It is very important that appropriate methods be used for analyzing the size distribution of biomass particles. Rezaei et al. [49] showed that biomass particles have an interaction with the sieves and thus mechanical sieving underestimates the size distribution of the ground biomass particles. It seems that digital techniques such as laser diffraction size analyzers are preferred options.

2.3.10 A CASE STUDY – EFFECT OF MASS AND VOLUME SHRINKAGE ON DRYING RATE [52]

In this case study at the University of British Columbia (UBC), Rezaei et al. [52] showed that particle density and shape influence the shrinkage of either single particles or their bulk. In this study, the ground chip and ground pellet particles with various initial moisture contents were dried and their total particle volume shrinkage coefficients were calculated using Equation (2.12). The average particle size of particles was measured using a digital particle size analyzer (Malvern Mastersizer, dry module, model 2000).

Figure 2.19 shows the reduction in average particle size of the ground chip and ground pellet particles as their moisture content drops. From an initial moisture content of 0.90 down to 0.50 (kg water/kg dry mass basis), a slight reduction in average particle diameter (d_p) happened. Below the moisture content of 0.50 kg water/kg dry mass basis, the dp of both chip and pellet particles dropped sharply. It seems that particles were completely saturated with water at an initial moisture content of 0.50 kg water/kg dry mass basis (and higher). The variation of dp with moisture content was fitted to a proposed power-law empirical correlation.

$$d_p = a.M^b \tag{2.13}$$

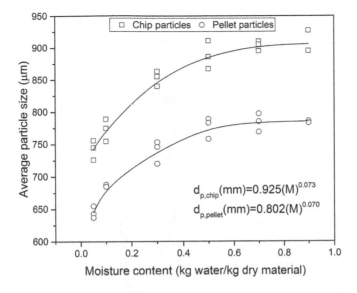

FIGURE 2.19   Variation of average particle size of samples as a function of moisture content [52].

The acquired data in Figure 2.19 were converted to the shrinkage coefficients of particle volume using Equation (2.12). Furthermore, the volume reduction in the bulk of particles was measured before and after drying for biomass particles with various initial moisture contents. Figure 2.20 shows the total shrinkage in particle volume and bulk volume of chip and pellet particles at various initial moisture contents.

Shrinkage in both particle and bulk volume increased with initial moisture content and leveled off at an initial moisture content of 0.50 kg water/kg dry mass basis. It means that the particles are fully saturated with water with initial moisture beyond 0.50 kg water/kg dry material. Pellet particles exhibited a larger shrinkage that ground chip particles. The reason is the difference in the microscopic structure of the chip and pellet particles. Rezaei et al. [49] showed that pellet particles are more spherical than ground chip particles, which have a long and needle-like shape. Pellet particles also have a random orientation, so that they shrink similarly in all directions. Chip particles mostly shrink in thickness.

As explained earlier, the moisture diffusion coefficient was calculated using the fitting of experimental data in the falling rate period with the analytical solution of Fick's second law. Based on the assumption that the moisture diffusion coefficient does not depend on the moisture content in the falling drying rate period, the first term of a solution of Fick's second law is linearized and reduced to Equation (2.14).

$$\ln(MR) = \ln(A) - \left(\pi^2 D / R_p^2\right).t \qquad (2.14)$$

The experimental drying data were analyzed using Equation (2.14) to estimate the moisture diffusion coefficient in two cases. Case 1 is when the particle dimension is fixed during drying and case 2 is when the particle shrinks and its dimension changes. The independent variable is t for the former case and t/d_p^2 for the latter case,

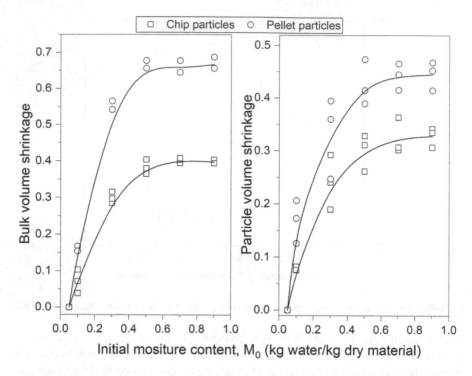

FIGURE 2.20 Total shrinkage in bulk volume and particle volume of ground chip and ground pellet particles due to drying [52].

respectively. The slope of the linearized equation is $-4\pi 2D/d_p^2$ for the former case and $-4\pi 2D$ for the latter case, respectively.

Table 2.5 lists the estimated parameters of D and A using the data presented in Figures 2.19 and 2.20. The fitting evaluation coefficients (R^2, χ^2, and RMSE) are also given in Table 2.5. The value of coefficient A was supposed to be close to unity since the MR = 1 at the beginning of the process ($t = 0$). Knowing this fact and looking at fitting evaluation coefficients (R^2, χ^2, and RMSE) shows that the first order drying equation with variable radius fitted the experimental data better than the equation with a fixed radius. The estimated moisture diffusion coefficients for the chip and pellet particles were $7.69–9.09 \times 10^{-11}$ m²/s and $3.13–4.59 \times 10^{-11}$ m²/s, respectively. This confirms that the dense structure of pellet particles reduces the rate of moisture diffusion compared to chip particles. Moisture diffusivity in the case of the fixed radius was a little higher than that in the case of variable radius. It is concluded that shrinkage during a drying process reduces moisture diffusion coefficients. A similar conclusion was drawn for drying of agricultural materials by Fusco et al. [71]. No specific trend in D with increasing initial moisture content was detected.

2.4 CONCLUSIONS

The current chapter focuses on drying of biomass feedstock to be used in combustion-based power plants; especially, when the intent is to replace coal with biomass

TABLE 2.5
Estimated Diffusion Coefficient (D, m²/s) and Constant A [52]

Calculation Mode	Parameters	M_0 (kg water/kg dry material)		
		0.10	0.30	0.50
	Ground Chip Particles			
Fixed radius	D (m²/s)	1.39E–10	1.87E–10	1.46E–10
	A	1.76	3.92	2.83
	d_p (mm)	0.773	0.853	0.888
	R^2	0.974	0.961	0.983
	χ^2	0.018	0.050	0.0150
	RMSE	0.135	0.225	0.123
Variable radius	D (m²/s)	8.33E–11	9.09E–11	7.69E–11
	A	1.09	1.06	1.14
	d_p (mm)		$0.925(M)^{0.073}$	
	R^2	0.989	0.998	0.994
	χ^2	0.007	0.002	0.004
	RMSE	0.087	0.051	0.070
	Ground Pellet Particles			
Fixed radius	D (m²/s)	6.13E–11	7.6E–11	9.21E–11
	A	1.657	1.963	3.950
	d_p (mm)	0.686	0.740	0.777
	R^2	0.973	0.973	0.973
	χ^2	0.033	0.036	0.038
	RMSE	0.181	0.189	0.196
Variable radius	D (m²/s)	3.13E–11	4.59E–11	4.47E–11
	A	1.03	1.05	1.17
	dp (mm)		$0.802(M)^{0.070}$	
	R^2	0.984	0.992	0.999
	χ^2	0.031	0.007	0.001
	RMSE	0.178	0.084	0.036

fuel in the existing facilities. Reviewing the open literature shows that a range of bio-mass particles' properties such as dimensions, shape, and density influence combustion efficiency. The combined effects of the biomass properties influence the kinetics and rate of combustion.

Either wood chip or wood pellet may be used as biomass feedstock for power generation applications. An important advantage of the wood pellet is its low moisture content, although wood pellets may be re-wetted due to poor storage/transportation and adsorb water or moisture. A review of the literature and industrial practice indicates that biomass should be dried down to around 30% prior to feeding a combustion chamber. Therefore, size reduction and drying are the essential pre-processing prior to feeding the fuel to the combustion chamber. In coal power plants, the biomass fuel particles are pulverized and flow pneumatically using the re-circulated hot gases in the pipelines feeding the combustion chamber. Using hot gas pneumatic transportation dries the pulverized fuel particles and also enhances the process' thermal efficiency.

A serious challenge for replacing coal with renewable biomass fuel in power generation stations is to modify the existing facilities to make them compatible with the new fuel. One of the most important facilities to be modified is the coal pulverizer. OPG published documents identifying key modifications of roll-race or ball-race coal pulverizers to crush wood fuel. The first is the limited size reduction. The coal pulverizer is not compatible with the fibrous nature of biomass materials. However, the grinding elements in a coal mill reduce the size of wood pellets back into their constituent dust, which is mostly particles smaller than 2 mm. That is the reason why wood pellets seem to be a preferred form of biomass fuel for combustion application. However, the limited size reduction of white pellets still poses a significant problem for pneumatic transport and combustion, and more effort and process tuning are happening in this field. Another important modification is using cold primary air. In spite of hot pneumatic transport of lignite using the exhaust gas at the 280–300°C temperature regime, cold air should be used (<180°C) for biomass fuel to prevent pre-combustion of biomass fuel in the pipelines. The information obtained from Atikokan power generation station indicates that the air temperatures in the mill inlet and outlet are 140–160°C and 80–85°C regimes, respectively. Although temperature selection is more about thermal efficiency at the burner tip, the drying of biomass fuel also happens and helps to achieve a more efficient burning process.

Section 2.3 of this chapter presents more extensive information on drying of pulverized/ground biomass particles as an essential pre-treatment for a combustion process. The current chapter covers the fundamentals and applicable methodologies for the drying of pulverized/ground biomass particles. The case studies presented through the chapter clarify the discussed analysis methods.

REFERENCES

1. Kasparbauer, R.D., *The Effects of Biomass Pretreatments on the Products of Fast Pyrolysis*, Mechanical Engineering, Iowa State University, Master of Science Dissertation, 2009.
2. Marshall, L. and Gaudry, D., *The Application of the Dedicated Milling Concept for 100% Wood Firing at Atikokan Generating Station.* Ontario Power Generation, Atikokan GS OPG, 2011.
3. Rezaei, H., "Physical and thermal characterization of ground wood chip and ground wood pellet particles", Chemical and Biological Engineering, University of British Columbia (UBC), PhD Dissertation, 2017.
4. Carlson, T.R., Tompsett, G.A., Conner, W.C., and Huber, G.W., "Aromatic production from catalytic fast pyrolysis of biomass-derived feedstocks". *Topics in Catalysis*, 2009, 52 (3), 241–252.
5. Biagini, E., Narducci, P., and Tognotti, L., "Size and structural characterization of lignin-cellulosic fuels after the rapid devolatilization". *Fuel*, 2008, 87 (2), 177–186.
6. Lu, H., Ip, E., Scott, J., Foster, P., Vickers, M., and Baxter, L.L., "Effects of particle shape and size on devolatilization of biomass particle". *Fuel*, 2010, 89 (5), 1156–1168.

7. Zheng, A., Zhao, Z., Chang, S., Huang, Z., Wang, X., He, F., and Li, H., "Effect of torre-faction on structure and fast pyrolysis behavior of corncobs". *Bioresource Technology*, 2013, 128, 370–377.
8. Westover, T.L., Phanphanich, M., Clark, M.L., Rowe, S.R., Egan, S.E., Zacher, A.H., and Santosa, D., "Impact of thermal pretreatment on the fast pyrolysis conversion of southern pine". *Biofuels*, 2013, 4 (1), 45–61.
9. Deng, L., Zhang, T., and Che, D., "Effect of water washing on fuel properties, pyrolysis and combustion characteristics, and ash fusibility of biomass". *Fuel Processing Technology*, 2013, 106, 712–720.
10. Wang, Y., He, T., Liu, K., Wu, J., and Fang, Y., "From biomass to advanced bio-fuel by catalytic pyrolysis/hydro-processing: Hydrodeoxygenation of bio-oil derived from biomass catalytic pyrolysis". *Bioresource Technology*, 2012, 108, 280–284.
11. Zanzi, R., Sjöström, K., and Björnbom, E., "Rapid pyrolysis of straw at high tempera-ture", in *Developments in Thermochemical Biomass Conversion*, Bridgwater, A.V. and Boocock, D.G.B., editors. 1997, Springer Netherlands, pp. 61–66.
12. Roy, M.M., Dutta, A., and Corscadden, K., "An experimental study of combustion and emissions of biomass pellets in a prototype pellet furnace". *Applied Energy*, 2013, 108, 298–307.
13. Hupa, M., Karlström, O., and Vainio, E., "Biomass combustion technology develop-ment – It is all about chemical details". *Proceedings of the Combustion Institute*, 2017, 36 (1), 113–134.
14. Hecht, E.S., Shaddix, C.R., Molina, A., and Haynes, B.S., "Effect of CO_2 gasification reaction on oxy-combustion of pulverized coal char". *Proceedings of the Combustion Institute*, 2011, 33(2), 1699–1706.
15. Dasappa, S., Paul, P.J., Mukunda, H.S., and Shrinivasa, U., "Wood-char gasifica-tion: experiments and analysis on single particles and packed beds". *Symposium (International) on Combustion*, 1998, 27 (1), 1335–1342.
16. Rezaei, H., Sokhansanj, S., Bi, X., Lim, C.J., and Lau, A., "A numerical and experimen-tal study on fast pyrolysis of single woody biomass particles". *Applied Energy*, 2017, 198, 320–331.
17. Miccio, F., Barletta, D., and Poletto, M., "Flow properties and arching behavior of bio-mass particulate solids". *Powder Technology*, 2013, 235, 312–321.
18. Jensen, P.D., Temmerman, M., and Westborg, S., "Internal particle size distribution of biofuel pellets". *Fuel*, 2011, 90 (3), 980–986.
19. Tooyserkani, Z., Kumar, L., Sokhansanj, S., Saddler, J., Bi, X.T., Lim, C.J., Lau, A., and Melin, S., "SO_2-catalyzed steam pretreatment enhances the strength and stability of softwood pellets". *Bioresource Technology*, 2013, 130, 59–68.
20. Bridgwater, A.V. and Peacocke, G.V.C., "Fast pyrolysis processes for biomass". *Renewable and Sustainable Energy Reviews*, 2000, 4 (1), 1–73.
21. Demirbas, A., "Effects of temperature and particle size on bio-char yield from pyroly-sis of agricultural residues". *Journal of Analytical and Applied Pyrolysis*, 2004, 72 (2), 243–248.
22. Czernik, S. and Bridgwater, A.V., "Overview of applications of biomass fast pyrolysis oil". *Energy & Fuels*, 2004, 18 (2), 590–598.
23. Bridgwater, A.V., "Review of fast pyrolysis of biomass and product upgrading". *Biomass and Bioenergy*, 2012, 38, 68–94.
24. Isahak, W.N.R.W., Hisham, M.W.M., Yarmo, M.A., and Yun Hin, T.-y., "A review on bio-oil production from biomass by using pyrolysis method". *Renewable and Sustainable Energy Reviews*, 2012, 16 (8), 5910–5923.
25. NikAzar, M., Hajaligol, M.R., Sohrabi, M., and Dabir, B., "Effects of heating rate and particle size on the products yields from rapid pyrolysis of beech-wood". *Fuel Science & Technology International*, 1996, 14 (4), 479–502.

26. Ball, R., C. McIntosh, and Brindley, J. "The role of char-forming processes in the thermal decomposition of cellulose". *Physical Chemistry Chemical Physics*, 1999, 1 (21), 5035–5043.
27. Di Blasi, C., "Modeling intra- and extra-particle processes of wood fast pyrolysis". *AIChE Journal*, 2002, 48 (10), 2386–2397.
28. Grønli, M.G. and Melaaen, M.C., "Mathematical model for wood pyrolysis – comparison of experimental measurements with model predictions". *Energy & Fuels*, 2000, 14 (4), 791–800.
29. Angın, D., "Effect of pyrolysis temperature and heating rate on biochar obtained from pyrolysis of safflower seed press cake". *Bioresource Technology*, 2013, 128, 593–597.
30. Altun, N.E., Hicyilmaz, C., and Kök, M.V., "Effect of particle size and heating rate on the pyrolysis of Silopi asphaltite". *Journal of Analytical and Applied Pyrolysis*, 2003, 67 (2), 369–379.
31. Peng, J.H., Bi, H.T., Sokhansanj, S., and Lim, J.C., "A study of particle size effect on biomass torrefaction and densification". *Energy & Fuels*, 2012, 26 (6), 3826–3839.
32. Rezaei, H., Yazdanpanah, F., Lim, C.J., Lau, A., and Sokhansanj, S., "Pyrolysis of ground pine chip and ground pellet particles". *The Canadian Journal of Chemical Engineering*, 2016, 94 (10), 1863–1871.
33. Vamvuka, D., Bio-Oil, "Solid and gaseous biofuels from biomass pyrolysis processes—an overview". *International Journal of Energy Research*, 2011, 35 (10), 835–862.
34. Uslu, A., Faaij, A.P.C., and Bergman, P.C.A., "Pre-treatment technologies, and their effect on international bioenergy supply chain logistics. Techno-economic evaluation of torrefaction, fast pyrolysis and pelletisation". *Energy*, 2008, 33 (8), 1206–1223.
35. Butler, E., Devlin, G., Meier, D., and McDonnell, K., "A review of recent laboratory research and commercial developments in fast pyrolysis and upgrading". *Renewable and Sustainable Energy Reviews*, 2011, 15 (8), 4171–4186.
36. Pang, S. and Mujumdar, A.S., "Drying of woody biomass for bioenergy: Drying technologies and optimization for an integrated bioenergy plant". *Drying Technology*, 2010, 28 (5), 690–701.
37. Scott, D.S. and Piskorz, J., "The continuous flash pyrolysis of biomass". *Canadian Journal of Chemical Engineering*, 1984, 62 (3), 404–412.
38. Baxter, L., "Biomass-coal co-combustion: Opportunity for affordable renewable energy". *Fuel*, 2005, 84 (10), 1295–1302.
39. Lam, P.S., Sokhansanj, S., Bi, X., Lim, C.J., Naimi, L.J., Hoque, M., Mani, S., and Womac, A.R., "Bulk density of wet and dry wheat straw and switch grass particles". *Applied Engineering in Agriculture*, 2008, 24 (3), 351–358.
40. Chen, D., Zhang, Y., and Zhu, X., "Drying kinetics of rice straw under isothermal and nonisothermal conditions: A comparative study by thermogravimetric analysis". *Energy & Fuels*, 2012, 26 (7), 4189–4194.
41. Chen, D., Zheng, Y., and Zhu, X., "Determination of effective moisture diffusivity and drying kinetics for poplar sawdust by thermogravimetric analysis under isothermal condition". *Bioresource Technology*, 2012, 107, 451–455.
42. Chen, D.-Y., Zhang, D., and Zhu, X.-F., "Heat/mass transfer characteristics and non-isothermal drying kinetics at the first stage of biomass pyrolysis". *Journal of Thermal Analysis and Calorimetry*, 2012, 109 (2), 847–854.
43. Berberović, A. and Milota, M.R., "Impact of wood variability on the drying rate at different moisture content levels". *Forest Products Journal*, 2011, 61 (6), 435–442.
44. Dedic, A., "Convective heat and mass transfer in moisture desorption of oak wood by introducing characteristic transfer coefficients". *Drying Technology*, 2000, 18 (7) 1617–1627.
45. Dorri, B., Emery, A.F., and Malte, P.C., "Drying rate of wood particles with longitudinal mass transfer". *Journal of Heat Transfer*, 1985, 107 (1), 12–18.

46. Tannous, K., Lam, P.S., Sokhansanj, S., and Grace, J.R., "Physical properties for flow characterization of ground biomass from Douglas fir wood". *Particulate Science and Technology*, 2012, 31 (3), 291–300.
47. Erlich, C., Björnbom, E., Bolado, D., Giner, M., and Fransson, T.H., "Pyrolysis and gasification of pellets from sugar cane bagasse and wood". *Fuel*, 2006, 85 (10–11), 1535–1540.
48. Tumuluru, J.S., Sokhansanj, S., Lim, C.J., Bi, T., Lau, A., Melin, S., Sowlati, T., and Oveisi, E., "Quality of wood pellets produced in british columbia for export". *Applied Engineering in Agriculture*, 2010, 26 (6), 1013–1020.
49. Rezaei, H., Lim, C.J., Lau, A., and Sokhansanj, S., "Size, shape and flow characterization of ground wood chip and ground wood pellet particles". *Powder Technology*, 2016, 301, 737–746.
50. Paulauskas, R., Džiugys, A., and Striūgas, N., "Experimental investigation of wood pellet swelling and shrinking during pyrolysis". *Fuel*, 2015, 142, 145–151.
51. Rezaei, H., Lim, C.J., Lau, A., Bi, X., and Sokhansanj, S., "Development of empirical drying correlations for ground wood chip and ground wood pellet particles". *Drying Technology*, 2017, 35 (12), 1423–1432.
52. Rezaei, H., Sokhansanj, S., Lim, C.J., Lau, A., and Bi, X., "Effects of the mass and volume shrinkage of ground chip and pellet particles on drying rates". *Particuology*, 2018, 38, 1–9.
53. Myllymaa, T., Holmberg, H., Hillamo, H., Laajalehto, T., and Ahtila, P., "Wood chip drying in fixed beds: Drying kinetics and economics of drying at a municipal combined heat and power plant site". *Drying Technology*, 2015, 33 (2), 205–215.
54. Tooyserkani, Z., Sokhansanj, S., Bi, X., Lim, J., Lau, A., Saddler, J., Kumar, L., Lam, P.S., and Melin, S., "Steam treatment of four softwood species and bark to produce torrefied wood". *Applied Energy*, 2013, 103, 514–521.
55. Orang, N. and Tran, H., "Effect of feedstock moisture content on biomass boiler operation". *Tappi J*, 2015, 14, 629–637.
56. Zhu, Z. and Kaliske, M., "Numerical simulation of coupled heat and mass transfer in wood dried at high temperature". *Heat and Mass Transfer*, 2011, 47 (3), 351–358.
57. Bahadori, A., Zahedi, G., Zendehboudi, S., and Jamili, A., "Estimation of the effect of biomass moisture content on the direct combustion of sugarcane bagasse in boilers". *International Journal of Sustainable Energy*, 2014, 33 (2), 349–356.
58. Gómez-de la Cruz, F.J., Cruz-Peragón, F., Casanova-Peláez, P.J., and Palomar-Carnicero, J.M., "A vital stage in the large-scale production of biofuels from spent coffee grounds: The drying kinetics". *Fuel Processing Technology*, 2015, 130, 188–196.
59. Wan Nadhari, W.N.A., Hashim, R., Danish, M., Sulaiman, O., and Hiziroglu, S., "A model of drying kinetics of Acacia mangium wood at different temperatures". *Drying Technology*, 2014, 32 (3), 361–370.
60. Sigge, G.O., Hansmann, C.F., and Joubert, E., "Effect of temperature and relative humidity on the drying rates and drying times of green bell peppers (*Capsicum annuum*)". *Drying Technology*, 1998, 16(8), 1703–1714.
61. Tapia-Blácido, D., Sobral, P.J., and Menegalli, F.C., "Effects of drying temperature and relative humidity on the mechanical properties of amaranth flour films plasticized with glycerol". *Brazilian Journal of Chemical Engineering*, 2005, 22, 249–256.
62. Mujumdar, A.S., *Handbook of Industrial Drying*, 2006, CRC Press, Tylor & Francis.
63. Rezaei, H. and Sokhansanj, S., "Physical and thermal characterization of ground bark and ground wood particles". *Renewable Energy*, 2018, 129 (Part A), 583–590.
64. Almeida, G., Assor, C., and Perré, P., "The dynamic of shrinkage/moisture content behavior determined during drying of microsamples for different kinds of wood". *Drying Technology*, 2008, 26 (9), 1118–1124.

65. Mazzanti, P., Togni, M., and Uzielli, L., "Drying shrinkage and mechanical properties of poplar wood (*Populus alba* L.) across the grain". *Journal of Cultural Heritage*, 2012, 13 (3Supplement), S85–S89.

66. Sturm, B., Nunez Vega, A.-M., and Hofacker, W.C., "Influence of process control strategies on drying kinetics, colour and shrinkage of air dried apples". *Applied Thermal Engineering*, 2014, 62 (2), 455–460.

67. Bennamoun, L. and Belhamri, A., "Mathematical description of heat and mass transfer during deep bed drying: Effect of product shrinkage on bed porosity". *Applied Thermal Engineering*, 2008, 28 (17–18), 2236–2244.

68. Kowalski, S.J. and Mierzwa, D., "Numerical analysis of drying kinetics for shrinkable products such as fruits and vegetables". *Journal of Food Engineering*, 2013, 114 (4), 522–529.

69. Thakor, N.J., Sokhansanj, S., Sosulski, F.W., and Yannacopoulos, S., "Mass and dimensional changes of single canola kernels during drying". *Journal of Food Engineering*, 1999, 40 (3), 153–160.

70. Sokhansanj, S., Bailey, W.G., and van Dalfsen, K.B., "Drying of North American ginseng roots (*Panax quinquefolius* L.)". *Drying Technology*, 1999, 17 (10), 2293–2308.

71. Fusco, A.J., Avanza, J.R., Aguerre, R.J., and Gabitto, J.F., "A diffusional model for drying with volume change". *Drying Technology*, 1991, 9 (2), 397–417.

3 Advances in Biodrying of Sludge

César Huiliñir Curío, Francisco Stegmaier, and Silvio Montalvo

CONTENTS

3.1 INTRODUCTION: WHAT IS BIODRYING?

Biological drying, or simply "biodrying," is an alternative pretreatment method aimed at combustion that has been developed in recent years (Zhao et al., 2010). Biodrying, which is based on a process similar to composting, aims to remove water from biowastes with high water content using the heat generated during the aerobic degradation of organic substances, in addition to forced aeration (Navaee-Ardeh et al., 2006). Biodrying of sludge can also (in contrast with landfilling) reduce fossil fuel requirements and consequently greenhouse gas emissions if combusted to produce steam and or power, thereby contributing positively to prevent climate change (Winkler et al., 2013).

Drying has long been recognised as part of the composting process (Haug, 1993), and it is usually viewed as a secondary and unwanted effect of this process. The first efforts in order to use the energy of aerobic oxidation for drying wet organic waste were made by Baader et al. (1975), who dried swine and dairy wastes. Later, Jewell et al. (1984), dried dairy manure and employed for the first time the term "biodrying" for the process of using the exothermic energy coming from the degradation of organic matter by microorganisms. Since these two pioneering works, several researchers have developed and optimised biodrying of different kinds of wastes,

focusing mainly on municipal solid wastes (MSW). The biodrying of sludge begins
with Frei et al. (2004), who worked with sewage sludge coming from pulp and paper
wastewater treatment intended for combustion in wood-waste furnaces. Later, the
same research group worked in continuous and batch systems in order to optimise
the process.

According to this background, biodrying can be defined as the process where
microorganisms aerobically degrade the organic matter in the wastes, a process
in which metabolic heat is produced, converting the liquid water contained in the
organic wastes into water vapour which is then transported and removed by airflow
(Navaee-Ardeh et al., 2006; Velis et al., 2009; Yang et al., 2017). Therefore, metabolic heat production, air convection, and molecular diffusion of oxygen and water
vapour are the main mechanisms involved in water removal from wet wastes by
biodrying (Frei et al., 2004). A scheme of the process is shown in Figure 3.1.

The biodrying process is aimed at producing a high quality solid recovered fuel
(SRF). This is achieved by: (1) increasing the energy content (EC) (Adani et al.,
2002) by maximizing the removal of moisture present in the waste matrix and preserving most of the gross calorific value of the organic chemical compounds through
minimal biodegradation; (2) facilitating the incorporation of the partly preserved bio-
genic content into the SRF; and (3) rendering the output more suitable for mechanical processing by reducing its adhesiveness (Velis et al., 2009). During this process
the organic matrix is both a substrate for microorganisms (which produce heat for
drying) and the end product (fuel/granules), which has a high energy value and can
be used as a replacement for coal and for thermal energy generation. This principle

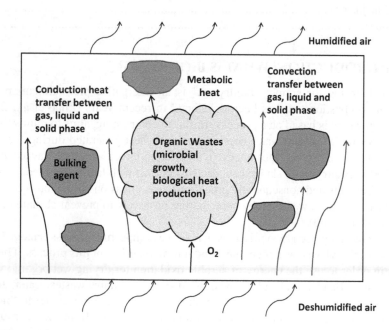

FIGURE 3.1 Simplified schematic of the biodrying process (modified from Yang et al.,
2017).

has been described for municipal waste and is often referred to as mechanical bio-logical treatment (MBT) (Ofori-Boateng et al., 2013; Velis et al., 2009), but it is not well recognised for excess sludge treatment. Biodrying of sludge can (in contrast to landfilling) reduce fossil fuel requirements and therefore greenhouse gas emissions if combusted to produce steam and or power, thereby positively contributing to pre-vent climate change (Navaee-Ardeh et al., 2006; Rada et al., 2009).

The present chapter will review the advances in the biodrying of sewage sludge. A first step will be to define the characteristics of sewage sludge as a possible solid fuel after a drying process, including a brief explanation of the difference between biodrying of sewage sludge and other moist wastes such as municipal solid wastes. Later, a deep review of the literature will be made and the advances in modelling the process will also be addressed.

3.2   SEWAGE SLUDGE AND ITS CHARACTERISTICS FOR THE BIODRYING PROCESS

Sewage sludge is the by-product of modern wastewater treatment plants (WWTP), generated during primary, secondary, or advanced treatment of municipal wastewa-ter. In 1988 the United States Environmental Protection Agency (EPA) estimated an annual domestic output of 5.4 million metric tons of sludge (Liang et al., 2003). In European countries, average sludge production is about 0.090 kg of dry sludge per person per day (Chen et al., 2002). Since raw sludge contains more than 90% water, the volume of the raw sludge produced is enormous. Thus, the emphasis in sludge management has shifted from its disposal to the beneficial use of it and its products. Recently, the utilisation of dried sewage sludge as a solid fuel has attracted much attention (Frei et al., 2004). However, due to the high moisture content (MC; typi-cally around 80 wt%) in dewatered sewage sludge, the development of an economi-cally viable drying process is still a challenge.

Sludge contains various physical states of water, including free water, intersti-tial water, surface water, intracellular water, and chemically bound water (Yang et al., 2014). The water that can be removed by mechanical dewatering is usually termed free water and includes the truly free, interstitial, and partial surface moisture. The remaining water is termed bound water, which includes the intra-cellular and chemically bound water together with the partial surface moisture (Chen et al., 2002). Conventional dewatering processes such as filter presses, belt presses, centrifugation, etc., can only remove the free water, but the remain-ing surface and intracellular (bound) water require more extensive treatment techniques for its removal (Navaee-Ardeh et al., 2006). The release of internal water held inside the cell's structure, which accounts for 70–80% of the packed cell mass, requires the disruption of sludge cells (Yang et al., 2014). Removal of the chemically bound water and the remaining surface moisture requires a phase change to water vapour by heat. Several techniques have been proposed for drying sewage sludge and their advantages and disadvantages are shown in Table 3.1.

The main advantage of the biodrying process is the self-generated energy for the drying process through a biochemical reaction. This advantage depends on the

TABLE 3.1

Advantages and Disadvantages of Most Used Conventional Drying Technologies

Technology	Advantages	Disadvantages
Rotary dryer	• High degree of product dryness • Provides for drying and granulation in one step • Allows for reclamation of waste heat • Versatile in configuration	• Large floor space requirement • High capital and maintenance costs • Back-mixing to avoid sludge agglomeration required
Paddle dryers	• Well-suited for heavy products • High ratio of heat transfer surface area to overall dryer volumen • Self-cleaning of the intermeshing paddles • Low space requirement	• Relatively high complexity of equipment • High maintenance costs • Sensitivity to large agglomerates, grit, or rocks in the sludge • Jamming or premature wear of the paddles
Belt dryers	• 80–90% energy recovery from the drying process possible • Closed-loop steam cycle • Ability for low-grade energy usage • No back-mixing of dried product with wet feed required • Low air temperatures required (30–90°C)	• Large floor space requirement • Relatively high degree of auxiliary equipment • High complexity of equipment
Superheated steam pneumatic dryer	• Well-proven, commercially available technology • Compact design • Short product residence time • Low floor space requirements	• Back-mixing of dried product with wet feed • Relatively high specific energy consumption • Turbulent drying conditions and high capital cost
Solar dryers	• Low energy requirements • Low maintenance costs • Environmentally sustainable	• Climate-related efficiency • Large space requirement • Odour problems • Slower drying rate in comparison to conventional dryers
Biodrying	• Low energy requirements • Low maintenance costs • Environmentally sustainable	• Odour problems • Slower drying rate in comparison to conventional dryers

biodegradable content as well as the water content (moisture content) of the sewage sludge that supports the microbial activity. Thus, the biodrying process has a natural limit, which is the minimum moisture content that allows the microbial activity. From composting studies, it was shown that below 20 wt% of water very little or no microbial activity occurs (Haug, 1993). For the biodrying processes, the minimum MC below which the biodegradation process is inhibited has not been well identified.

Until now, an MC of 48.9 wt% is probably the lowest for sludge biodrying (Yang et al., 2017).

Because of the need to reuse the huge amount of sewage sludge around the world, its proposed use as solid fuel has a sustainable solution. In this sense, the dry matter content or lower heating value (LHV) and composition of sludge are important factors for using it as a fuel. The composition of digested sludge is about 67% carbon, 5% hydrogen, 25% oxygen, 2.2% nitrogen, 0.8% sulphur (Stasta et al., 2006). Dried digested sludge from municipal WWTP has a composition similar to that of brown coal, but the LHV of digested sludge is lower (8–12 MJ/kg, compared to 21 MJ/kg for brown coal) (Stasta et al., 2006). Furthermore, the LHV depends mainly on the MC of sludge as shown in Figure 3.2, where the original MC value of sewage sludge has an LHV lower than 1 MJ/kg, while a sewage sludge with MC 50% has an LHV of around 5 MJ/kg. Values of MC obtained in biodrying processes varied widely, depending on the conditions used. According to Yang et al. (2017), the final products of the biodrying processes contain more than 55 wt% moisture (Frei et al., 2004; Huiliñir and Villegas, 2015; Zhao et al., 2010), although values as low as 33% have been reported (Winkler et al., 2013). An MC of around 55 wt% is too wet for its direct combustion, because the critical level of MC of combustible materials should be lower than 55 wt% to guarantee stable conditions in the boiler (Navaee-Ardeh et al., 2011b; Yang et al., 2017). Therefore, normally wet biodried wastes are mixed with dry fuels, so the wetness of biodried solids may not be a problem (Rulkens, 2008; Stasta et al., 2006). In the case of the sole combustion of biodried products, however, less heat would be produced due to this extra water. In any case, lowering the MC of biodried solids is crucial for combustion.

Because biodrying has been applied mainly to MSW, a comparison will be made with respect to this waste. There are two main differences between the biodrying of MSW and of sewage sludge (SS): (a) organic matter content and (b) initial moisture content. Regarding organic matter content, it is known that the biomass energy of sludge available for heat generation is limited due to low biodegradable

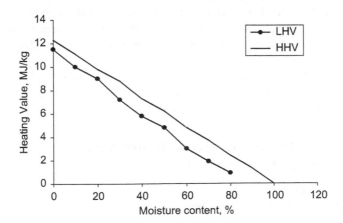

FIGURE 3.2 Relation between lower heating value (LHV), higher heating value (HHV), and water content (modified from Stasta et al. (2006)).

VS (Zhao et al., 2010), as most of the organics are separated by microbial cell membranes and unavailable for biodegradation (Ma et al., 2016). This characteristic also affects the bulking agent needed to give porosity to the sewage sludge, because the bulking agent can help to increase not only the porosity, but also the organic matter content. With respect to initial MC, the high initial MC in dewatered sludge (typically around 80 wt%) hinders the biodrying process because the excess moisture in the sludge causes packing and void space reduction, preventing proper air movement throughout the matrix (Richard et al., 2002; Yang et al., 2014), a situation that does not appear in the process with MSW. High MC, lack of porosity, and compacting tendency make dewatered sludge somewhat unique and difficult to deal with (Zhao et al., 2010). In this context, the use of a bulking agent (BA) increases the porosity of the biomass matrix, improving the effective aeration. Furthermore, the BA allows for the adjustment of the moisture content, N-content, and C/N ratio. Finally, the BA needs to be properly selected and used in optimum amounts to minimise its purchasing and transportation costs. These reasons make the use of BA an important factor for the development of the sewage sludge biodrying process.

3.3   ADVANCES IN SEWAGE SLUDGE BIODRYING

Biodrying of sewage sludge and its modelling have been extensively reported in the literature during the last decade. These studies normally report not only the performance of biodrying, but also parameters and conditions that can improve the process. According to the literature, the main parameters that must be considered for moisture removal are the following:

1. **Amount of bulking agent (BA):** The bulking agent improves the porosity of the biomass matrix, but decreases biological activity and bioheat (Larsen and McCartney, 2000; Zhao et al., 2011). Therefore, the reduction in the rate of biological activity must be balanced against the increased porosity that the bulking agent favours (Frei et al., 2004).
2. **Inlet airflow:** Airflow is necessary to produce heat through the biological reaction and to remove water from the matrix, whose temperature affects the temperature of the airflow and consequently its water holding capacity (Zhao et al., 2010).
3. **Sludge moisture content:** This parameter is related to the biological activity. Minimum moisture content between 65 and 50% is necessary for an efficient biological activity (Chang and Chen, 2010).

3.3.1   EFFECT OF TYPE AND AMOUNT OF BULKING AGENT (BA) ON THE BIODRYING PROCESS

An aerobic degradation process such as composting or biodrying relies heavily on the type and quantity of bulking agent used (Adhikari et al., 2009). In the compost process, for example, the BA absorbs part of the leachate produced during the decomposition process, to keep the mixture moist and sustain a high microbial activity. Most bulking agents will act as buffers against the organic acids produced during the early stages of composting, thereby helping to maintain the mixture's pH within a range of 6–8 (Adhikari et al., 2009). The BA also gives structure and porosity to the mixture for proper aeration (Barrington et al., 2002) and should be well selected and used in optimum amounts to minimise its purchasing and transportation costs. Indeed, Kulcu and Yaldiz (2007), studying the compost of goats, showed that increasing the BA results in an improvement in uniform air dispersion in the matrix and decreased loss of organic matter.

The amount of BA also corrects the formula's moisture content and C/N ratio (Adhikari et al., 2009). The sewage sludge contains less biodegradable organic matter than other wastes such as municipal wastes, and therefore the use of bulking agents as wood chips, wheat straw, and rice husk with high C/N ratio (Adhikari et al., 2008; Chang and Chen, 2010) can improve the biological activity (Zhao et al., 2011). However, Larsen and McCartney (2000) found that bioheat has a negative correlation with the C/N ratio. An out of range C/N ratio (<15 or >30) can decrease the success of the biodrying process. Therefore, the reduction in the rate of biological activity and the bioheat must be balanced against the increased porosity that the bulking agent provides.

Different BA can produce different effects over an aerobic degradation process. In the composting process it is well known that different bulking agents not only modify the physical properties of the biomass matrix, but also change the biodegradation kinetics and aerobic degradation performance. Pasda et al. (2005), using wood chips and rice husks for compost sewage sludge, showed that the use of wood chips as BA decreases the organic matter content and increases the pH value after 14 days of incubation, compared to the use of rice husks as BA. Das et al. (2003) showed that the use of wood bark and a synthetic plastic BA produces different oxygen concentrations within the matrix, with different effects on the biokinetics. Chang and Chen (2010), studying the composting of food wastes, also showed different temperatures, acidification time, and pH values using sawdust and rice husks as BA.

Despite the multiple studies on the effect of the amount of BA on the composting process (Adhikari et al., 2009; Gea et al., 2007; Kim et al., 2008), this has been scarcely studied in the literature for the biodrying process. Choi et al. (2001), drying poultry manure, showed that the use of sawdust as BA enhances moisture removal by 30% compared to biodrying without BA, while Frei et al. (2004), using wood chips as BA, could not show a clear tendency of biodrying efficiency with sewage/BA ratio, although they found an optimal sewage/BA ratio. On the other hand, Zhao et al. (2011) showed that the use of a more biodegradable bulking agent can improve the biogenerated heat in the biodrying process. Yang et al. (2014) studied the use of air-dried sludge, shredded rubber, and sawdust as bulking agents. In that work, the difference was noted in several points. For instance, the free air space (FAS) was very different among the assays, with the sawdust showing much higher FAS (95.3%) than air-dried sludge (73.5%) and rubber (71.1%). Regarding other parameters, the highest temperature (71°C) was found in biodrying with air-dried sludge, while the temperatures of sludge beds with rubber and sawdust only reached 41 and 44°C, respectively; this also affected the removal of biodegradable volatile solids (BVS) and water, with the highest removal of H_2O in the system with air-dried sludge (52.9% of the water initially present in the mixed sludge) and also a slightly higher BVS removal. These results showed that air-dried sludge was a better bulking agent than rubber and sawdust. Li et al. (2015) studied the biodrying of anaerobically digested sludge (ADS) using wheat residues (WR) as bulking agents, in order to improve the biodegradability of the ADS. They succeeded in reducing the moisture content from 67 to 50% in 18 days, while the volatile solids (VS) content decreased gradually from 65.92 to 51.62%. An interesting result of this work was the separation of the ADS and WR contributions to provide the energy required to dry. They indicated that the contribution of the ADS to the biogenerated heat during the biodrying process was only 13.99%, while the contribution of the WR to the biogenerated heat was 86.01% in the biodrying system. Therefore, it is very important to select a bulking agent that can also contribute to bioheat generation.

3.3.2   EFFECT OF INLET AIRFLOW RATE (AFR) ON THE BIODRYING PROCESS

Aeration rate is the main operational variable for process control in biodrying (Adani et al., 2002; Sugni et al., 2005; Zawadzka et al., 2010). Inlet airflow rate can be manipulated to control matrix temperature, in turn affecting the air dew point and biodegradation kinetics (Velis et al., 2009). Most evidence indicates that comparatively effective heat removal can be achieved by higher rates, resulting in lower matrix temperature (Adani et al., 2002), with an optimal waste matrix temperature (T_{waste}) as low as 45°C. This decrease of T_{waste} allows a delay of biodegradation, decreasing the bio-stability of the final product, but increasing the final energy content. Therefore, a high airflow rate may be necessary for the production of a sufficiently high heating value (HV) of the dried biosolid, preserving most of the organic matter content.

Normally there are different ways for aerating the biodryer: continuous or intermittent; positive or negative. Among all these ways, most of the studies use intermittent and positive aeration, even at the industrial scale. Continuous aeration has the disadvantage of the increased cost of the process; however, it can also help to dry faster. The positive or negative aeration is related mainly to another problem in biodryers: uneven drying, which is produced because the air enters the matrix from the bottom, becoming wet and saturated with water vapour as it travels upward through the pile. Once the air is saturated, it can no longer absorb water molecules in the upper layers. As a result, the matrix is dried heterogeneously, being dryer in the bottom zone and moister in the upper zone (Velis et al., 2009; Yang et al., 2017). In order to improve this weakness, some strategies have been proposed. Frei et al. (2004) tested a sophisticated pipe with three holes and inverted airflow system for biodrying of a sewage sludge coming from the paper and pulp industry. This system used a central conduit either pumping or pulling air, while the other two pipes were on inverted airflow. The configuration was criticised for removing water from the wet portion and depositing it in the dry portion of the matrix, favouring biodegradation rather than biodrying (Navaee-Ardeh et al., 2006; Velis et al., 2009). Inverting the airflow led to a drop in the relative humidity of the outlet air, with a re-wetting of the matrix by the humid inlet air (Velis et al., 2009). Sugni et al., (2005), using MSW, experimented with a reactor that simulated daily inverted airflow by exchanging the positions of the upper and lower reactor layers. They saw a mitigation of the matrix temperature gradient and a more homogeneous content in terms of moisture compared to the unidirectional flow. However, this arrangement did not achieve early convergence with the ambient temperatures as in the unidirectional experiments, indicating the need for a prolonged residence time (Velis et al., 2009). As can be seen, a proposal that improves the uneven drying in this process is still lacking.

The effect of the inlet airflow rate (AFR) has been studied mainly using municipal solid wastes, with little research on sewage sludge. Adani et al. (2002), working with municipal solid waste, established that a high airflow rate is necessary for effective and fast biodrying, ensuring a high energy content (EC) in the final product. However, further studies with the same sample revealed low reproducibility of HV and EC, properties that are highly dependent on the laboratory conditions used to measure them (Sugni et al., 2005). Zawadzka et al. (2010), using the organic fraction of municipal solid wastes, showed that high AFR resulted in a high percentage of water removal; however, this did not contribute to high temperatures in the processed biomass, high heat of combustion values, or high HV. Zhao et al. (2010), using dewatered sewage sludge and a mixture of straw and sawdust as BA, showed that the increase on inlet AFR does not have a remarkable advantage regarding biodrying efficiency, especially with high turning frequency. Huiliñir and Villegas (2015) studied the simultaneous effect of MC and AFR on sewage sludge biodrying, indicating that initial MC has a stronger effect on biodrying than the AFR, affecting the air outlet temperature and improving water removal, with higher maximum temperatures obtained around 68% and the lowest maximum matrix temperature obtained at initial MC = 78%. An important effect observed by these researchers was the different temperatures of the matrix solid and the air surrounding it. According to the results, the difference between matrix temperature and outlet air temperature (ΔT_{m-a}) increases at lower AFR in all the assays. For instance, at an initial MC = 68%, the higher ΔT_{m-a} at AFR = 1 L/min kg total solids (TS) was 19°C, while at AFR = 2 L/min kg TS and AFR = 3 L/min kg TS was only 10°C and 2°C, respectively. This behaviour shows the effect of AFR on the energy transfer between the matrix and the air. At higher AFR there is a lower energy transfer resistance because of the decrease of the boundary layer on the solid matrix, allowing greater energy transfer and obtaining closer values between matrix and outlet air temperatures (Huiliñir and Villegas, 2015). This fact is also important in the modelling of the process.

Finally, the AFR also affects the cost of the process. It would be cost effective when airflow rates are moderate because the aeration rate determines the energy consumption of the process (Yang et al., 2017).

3.3.3 Effect of Moisture Content (MC) on the Biodrying Process

The MC of the waste matrix is the single most important variable not only for evaluating the performance of biodrying, but also for controlling the optimum initial conditions of the waste matrix (Velis et al., 2009). During aerobic biodegradation, around 0.5–0.6 g of metabolic water is produced per gram of BVS decomposed (Miller, 1989). However, water losses during biodrying are much greater than the metabolic water gains, resulting in a dried matrix (Richard et al., 2004). This MC reduction in biodrying may stop biodegradation, or its rate may be significantly reduced due to complete decomposition of readily biodegradable VS (degradation effect), or more probably, due to water stress where low moisture conditions inhibit microbial activity and movement (drying effect) (Miller, 1989). On the other hand, MC also affects the free air space (FAS), which is important in determining the quantity and movement of air through the matrix. Moisture levels should be high enough to assure adequate rates of biological reactions, yet not so high that FAS is eliminated, thus reducing the rate of oxygen transfer and, in turn, the rate of biological activity (Yang et al., 2014).

The optimum range for biodrying is between 60 and 65%, according to several studies. Yang et al. (2014) studied the effect of initial MC in the biodrying process. They showed that 50–70 wt% was the optimal initial MC range for the sludge bio-drying process, with VS reduction between 12.3 and 21.2%; however, they worked at only one constant airflow rate. Huiliñir and Villegas (2015) studied the simultaneous effect of initial moisture content and airflow rate on sewage sludge biodrying. They showed that the initial moisture content has a stronger effect on biodrying performance compared to the airflow rate (AFR), with initial MC values close to 60% which improve both water removal and VS biodegradation. Similar values were reported by Ma et al. (2016), who, using co-biodrying of sewage sludge and food waste, indicated that an initial moisture content of 62.68% presents the best biodrying results in terms of water removal rate and VS consumption.

MC also influences the dynamics of aerobic biodegradation. For example, during composting, optimum moisture conditions differ significantly, change during the process (either increase or decrease), and vary with the substrate (Richard et al., 2002) and the waste/bulking agent ratio (Tremier et al., 2009). Liang et al. (2003) used factorial design experiments to investigate the influence of temperature and MC on microbial activity, measured as O_2 uptake rate (mg g^{-1} h^{-1}) during composting of biosolids, showing that MC is more influential than temperature. The rate of heat production by microbial activity can be anticipated to decline as the MC of the matrix approaches the water stress limit, affecting the drying mechanism. From composting studies, it is evident that below 20% w/w very little or no microbial activity occurs (Velis et al., 2009). On the other hand, the water content of the initial sewage sludge over a reasonable range could make it possible to decrease the amount of bulking material required in the sewage sludge biodrying process. For instance, Li et al. (2015) indicated that for anaerobically digested sludge (ADS) and wheat residues (WR) as bulking agent, if the moisture content of the ADS is expected to decrease from 80 to 40%, the required weight of the WR is expected to be more than 0.28 kg (wet weight), and the ratio of ADS/WR is less than 3.59 (w/w, wet weight). Instead, when we need to decrease the initial moisture content from 60 to 40%, the required ADS/WR ratio is about 9.85 (w/w, wet weight) and the amount of bulking materials required is reduced considerably, also making the process economically more viable.

3.3.4 ENERGY AND MASS BALANCES IN THE PROCESS

The effectiveness of biodrying depends on the energy balance of the system, in which simultaneous heat and mass transfer are driven by biogenerated heat (He et al., 2013). Therefore, it is important to know, through mass and energy balances, how the energy coming from the biochemical reaction is used and how the water is distributed in the different streams of the process. Because this is a newly developed process, the research on energy balance in biodrying technology is still in the preliminary stage. However, since the energy input and output pathways in bio-drying systems are almost the same as those of composting, the analytical methods required for the energy balance model for the composting process can also be used for biodrying systems (He et al., 2013). Under these conditions, the boundary of a biodrying process system is basically the biodryer, where it is assumed that the bioreactor is a pure thermal balance model (He et al., 2013). According to this definition, the biodegradation of matter represents the heat source, and the heat loss pathway encompasses heat dispersion from ventilation, latent and sensible heat of evaporation, sensible heat of material heating, and the heat dissipation of the external wall of the reactor. A scheme of the process is shown in Figure 3.3.

The mass balance is related mainly to water. There is also a mass balance of CO_2; however, in order to see the most important aspects in terms of engineering, only the water balance will be analysed.

Because biodrying is a batch process in a non-steady state, the final water content is the result of the initial water content, water produced by metabolism, evaporated water, and water loss by turning in each time. This last term appears only when the matrix is turning. The basic equation of the integral water balance in the matrix, which is important for process analysis, at any time during sampling, can be written as follows (Huiliñir and Villegas, 2015):

$$W_{matrix,i} - W_{matrix,f} = W_{evap} + W_{turning} - W_{metabolic} \qquad (3.1)$$

where

$W_{matrix,i}$ is the initial water mass in the matrix;
$W_{matrix,f}$ is the final water mass in the matrix;
W_{evap} is the water mass lost by evaporation;
$W_{turning}$ is the water mass lost by turning,
and $W_{metabolic}$ is the generated water mass by microbial metabolism.

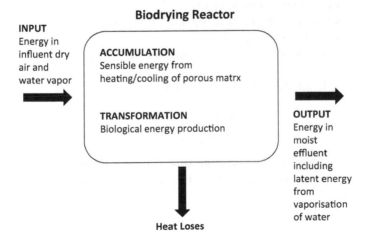

FIGURE 3.3 Boundaries for the mass and energy balance analysis (modified from He et al. 2013).

The initial ($W_{\text{matrix},i}$) and final water mass ($W_{\text{matrix},f}$) in the matrix are normally calculated by multiplying the mass in the bioreactor by its moisture content, measured as percentage, at the beginning and end of the process, respectively. The water mass generated by microbial metabolism ($W_{\text{metabolic}}$) is calculated assuming that water is generated only in the process of VS oxidation, where its value is:

$$W_{\text{metabolic}} = \text{VS}_{\text{consumed}} \cdot Y_{\text{H}_2\text{O/VS}} \tag{3.2}$$

where $Y_{\text{H}_2\text{O/BVS}}$ is a constant yield coefficient which is normally assumed as the average of the values obtained from the literature (Zhang et al., 2012). The water mass lost by evaporation (W_{evap}) is calculated using the inlet and outlet temperatures and relative humidity of the air. This way, the evaporated water can be calculated as:

$$W_{\text{evap}} = \int_{t_i}^{t_f} F_{\text{da}}(t) \cdot \left(Y_i - Y_o\right) dt \tag{3.3}$$

where

F_{da} is the inlet mass flow rate of dry air into the bioreactor and

Y_i and Y_o are the absolute humidity of the inlet and outlet air, respectively.

The dry air mass flow rate can be calculated as (Treybal, 1986):

$$F_{\text{da}}(t) = \cfrac{Q_{\text{air}}}{\left(\cfrac{1}{\text{MW}_{\text{air}}} + \cfrac{1}{\text{MW}_{\text{water}}}\right) \cdot R \cdot T_{\text{air,inlet}}(t)} \tag{3.4}$$

where

Q_{air} is the measured inlet airflow rate,

MW_{air} is the molecular mass of air,

MW_{water} is the molecular mass of water,

R is the universal gas constant.

and T is the temperature of the humid inlet air.

The specific humidity of air is calculated as:

$$Y = \frac{\text{MW}_{\text{water}}}{\text{MW}_{\text{air}}} \cdot \left(\frac{P_{v,w}}{P_{abs} - P_{v,w}}\right) \tag{3.5}$$

where $P_{v,w}$ is the actual water vapour pressure in air at a defined temperature and P_{abs} is the absolute gas pressure (assumed to be 1 atm). The actual vapour pressure can be calculated as:

$$P_{v,w} = \text{HR} \cdot P_{vs,w} \tag{3.6}$$

where $P_{vs,w}$ is the saturation water vapour pressure, which is dependent on temperature and can be mathematically modelled by the Antoine equation:

$$\log_{10} P_{vs,w} = 8.896 - \frac{2238}{T(^\circ K)} \tag{3.7}$$

Therefore, measuring the temperature and relative humidity of the inlet and outlet air, the W_{evap} can be calculated. Finally, the water mass loss by turning is calculated using the general balance (Equation 3.1):

$$W_{turning} = W_{matrix,i} - W_{matrix,f} - W_{evap} + W_{metabolic} \tag{3.8}$$

The energy balance can be calculated based on the methodology presented by Zhao et al. (2010) and by He et al. (2013). The energy balance equation indicated that the energy generated by metabolism is distributed between the energy for drying and lost energy, as follows:

$$Q_{bio} = Q_{consu} + Q_{loss} \tag{3.9}$$

where
Q_{bio} is the energy generated by bioreaction,
Q_{consu} is the energy consumed by the process,
and Q_{loss} is the energy lost in the process.

The consumed energy can be calculated as:

$$Q_{consum} = Q_{dryair} + Q_{watvap} + Q_{water} + Q_{solid} + Q_{evapo} + Q_{rad} + Q_{turning} \tag{3.10}$$

According to the last equations, the energy is consumed by aeration (Q_{dryair} and Q_{watvap}), by evaporated water (Q_{evapo}), by matrix temperature increase (Q_{water} and Q_{solid}), by radiation (Q_{rad}), and by turning ($Q_{turning}$). The following equations represent each one of these terms:

1. Biologically generated heat:

$$Q_{bio} = \Delta VS \cdot H_c \tag{3.11}$$

where ΔVS is the VS consumption and H_c is the biodegradation enthalpy.
2. Heat consumed by aeration (sensible heat):

$$Q_{dryair} = c_{p,dryair} \cdot \int_{t_i}^{t_f} F_{da}(t) \cdot (T_{air,outlet} - T_{air,inlet}) dt \tag{3.12}$$

$$Q_{watevap} = c_{p,watvap} \cdot \int_{t_i}^{t_f} (F_{da}(t) \cdot Y_o(t) \cdot T_{air,outlet}(t) - F_{da}(t) \cdot Y_i(t) \cdot T_{air,inlet}(t)) dt \tag{3.13}$$

3. Heat consumed by water evaporation (latent heat):

$$Q_{\text{evapo}} = \int_{t_i}^{t_f} F_{\text{da}}(t) \cdot \lambda_{\text{evap}}(t)dt \tag{3.14}$$

where λ_{evap} is the latent heat of water evaporation, which can be calculated as (Zhao et al., 2010):

$$\lambda_{\text{evap}}(t) = \left(1093.7 - 0.5683 \cdot \left(\frac{T_{\text{matrix}}(t) + 32}{5}\right) \cdot 9\right) \cdot \frac{1055}{454} \tag{3.15}$$

4. Heat consumed by matrix temperature increase (sensible heat):

$$Q_{\text{water}} = W_{\text{matrix}} \cdot c_{p,\text{water}} \cdot \Delta T_{\text{matrix}} \tag{3.16}$$

$$Q_{\text{solid}} = M_{\text{solid}} \cdot c_{p,\text{solid}} \cdot \Delta T_{\text{matrix}} \tag{3.17}$$

In this case, ΔT_{matrix} was the change of matrix temperature in a time element, i.e., between matrix water mass measurements.

5. Heat consumed by radiation:

$$Q_{\text{radi}} = \sigma \cdot A_{\text{top}} \cdot F_a \cdot F_e \cdot \left(T_{\text{air},0}^4 - T_a^4\right) \tag{3.18}$$

6. Heat consumed by turning:

$$Q_{\text{turning}} = W_{\text{matrix}} \cdot c_{p,\text{water}} \cdot \left(T_{\text{matrix}} - T_a\right) + M_{\text{solid}} \cdot c_{p,\text{solid}} \cdot \left(T_{\text{matrix}} - T_a\right) \tag{3.19}$$

With this global balance it is feasible to determine the energy lost in turning, as well as any of the energy terms used in this equation.

The energy balance shown earlier has been applied to several systems (Zhao et al., 2010; Huiliñir and Villegas, 2015; He et al., 2013) because of its simplicity.

3.4   ADVANCES IN THE MODELLING OF THE PROCESS

Limited modelling attempts for biodrying processes exist in the peer-review literature. These models use a deterministic approach and involve heat balance coupled with mass balance (Huiliñir et al., 2017; Navaee-Ardeh et al., 2011a,b; Pujol et al., 2011; Rada et al., 2007) or heat balance coupled with momentum and mass balance (Zambra et al., 2011). Rada et al. (2007) provided initial batch biodrying modelling results focused on the simulation of lower heating value (LHV) dynamics, volatile solids consumption, waste moisture content dynamics, and nitrogen compound release of municipal solid waste. Through relative humidity measurements in the air, they calculated the waste moisture content. However, they did not include equations for temperature prediction. Frei et al. (2004), using sewage sludge from the paper and pulp industry and wood bark as a bulking agent, succeeded in modelling the pneumatic behaviour matrix and the water mass balance of their complex inverted airflow configuration for batch biodrying; however, they did not show the heat balance and temperature predictions. Navaee-Ardeh et al. (2011b,c) presented a steady-state 1-D and 2-D model of a continuous biodrying process of sewage sludge without bulking agent, showing a good correlation between predicted temperatures and experimental data. Pujol et al. (2011) presented a windrow composting model taking into account the drying phenomena. They explored local equilibrium (LE) and local non-equilibrium (LNE) situations and concluded that the latter allow shorter calculation times and easier implementation. The main weakness of this work is the lack of validation of its results. Zambra et al. (2011) developed a 3-D mathematical model of a dynamic biodrying process, which was validated using the data of Rada et al. (2007). This model shows very good agreement with temperature data, but it does not present equations for water calculations or volatile solids consumption. De Guardia et al. (2012) presented work focused on the modelling of the heat transfers during composting in a pilot-scale reactor under forced aeration. Like in earlier works, this model calculates the temperature, but does not present results for dry solids. Recently, Huiliñir et al. (2017) presented a new model of the biodrying process, separating the solid and gas phases, including the VS degradation kinetics, and using local non-equilibrium (LNE). They also validated the model with experimental data, showing that the model is robust for the analysis of the systems under several operational conditions.

The models of the biodrying process that have been presented until now are supported mainly on composting models and involve heat and mass balance related to biokinetics, focusing on temperature prediction and calculating the water removal indirectly, i.e., supported on data (de Guardia et al., 2012; Navaee-Ardeh et al., 2011b; Pujol et al., 2011; Zambra et al., 2011). The majority of these works do not include the variations of dry solids explicitly in the model, with the exception of the work of Huiliñir et al. (2017). As was mentioned before, the BVS determination is very important in the biodrying process in order to determine the quality of biodrying and the increase of its heat value (Zawadzka et al., 2010). The BVS degradation is also connected with biological energy production (Mason, 2006). In this sense, the majority of the biodrying models (de Guardia et al., 2012; Navaee-Ardeh et al., 2011a,b; Zambra et al., 2011) have been based on the implicit approach, i.e., using oxygen consumption or CO_2 generation, and then applying appropriate heat yield factors in order to obtain an energy expression (Mason, 2006). Only Pujol et al. (2011) and Huiliñir et al. (2017) based the kinetics on the explicit form.

Regarding the moisture content, it has been simulated in composting and solid-state fermentation based on the assumption that there is moisture equilibrium between the bed and the air phase (de Guardia et al., 2012; Navaee-Ardeh et al., 2011a,b). For the moisture's local equilibrium, the gas vapour mass fraction is directly equal to the one at the liquid gas interface (Pujol et al., 2011). This will probably be approached if saturated air is used to aerate the bed; however, this situation is not a real condition in the biodrying process. In fact, data from batch (Rada et al., 2007; Zawadzka et al., 2010) and continuous (Navaee-Ardeh et al., 2010) processes show that the relative humidity of air (inlet and outlet) varies between 40 and 95%. Therefore, in this case the mathematical model must treat the substrate particles and inter-particle air as separate phases and describe the process of mass transfer of water between them, i.e., a local non-equilibrium approach should be used. This point was taken into account by Huiliñir et al. (2017), who proposed separating the matrix into two phases, the solid phase and the gas phase. This required the inclusion of a mass and energy transfer process between the solid and gas phases, using a local non-equilibrium approach. A scheme of the conceptual model is shown in Figure 3.4.

The mass and energy balances should be related to the biokinetics of the process, where energy for drying is produced. In the next chapter, the different kinetics models used in biodrying will be presented.

3.4.1 KINETICS OF SUBSTRATE DEGRADATION IN THE BIODRYING PROCESS

According to Hamelers (2004), two main strategies can be distinguished in the modelling of aerobic biokinetics: the inductive strategy, which is data based, and the deductive strategy, which is theory based. Typical examples of the inductive strategy are first-order substrate degradation kinetics and empirical substrate degradation kinetics in composting (Mason, 2006), and linear, exponential, logistic, and fast-acceleration/slow-deceleration kinetics in solid-state fermentation (Mitchell et al., 2004). For the theoretically based approach, the microbial population is taken explicitly into account mainly through Monod-type kinetics. This approach has been used in some papers in the composting process (Agamuthu et al., 2000; Lin et al., 2008; Pommier et al., 2008; Tremier et al., 2005; Zavala et al., 2004; Zhang et al., 2012), in solid-state fermentation (Figueroa-Montero et al., 2011; Mitchell et al., 2004), and

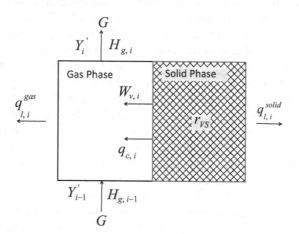

FIGURE 3.4 Conceptual model proposed by Huiliñir et al. (2017) in which G: dry air mass flow rate, $L_{dry\,air}$/min; Hg: enthalpy of gas phase (air), kJ; Y_i': absolute humidity of air at the gas outlet temperature in the i CSTR, kg H_2O/kg dry air; $q_{l,i}^{gas}$: energy transfer from the gas to the environment in the i layer, kJ/h; $q_{c,i}$: energy transfer by convection in the i layer, kJ/h; $q_{l,i}^{solid}$: energy transfer from the solid to the environment in the i layer, kJ/h; $W_{v,i}$: rate of water mass transfer, kg water/h; r_{VS}: reaction rate of the VS biodegradation.

recently also in the biodrying process (Huiliñir et al., 2017); however, the inductive strategy appears to be the most widely used and successful approach in composting, solid-state fermentation, and biodrying models (Huiliñir and Villegas, 2014; Mason, 2006; Mitchell et al., 2004; Villegas and Huiliñir, 2014). The main problem of the theoretical approach is the larger number of parameters compared to first-order kinetics (Hamelers, 2004). In the case of biodrying, the solid nature of the material and the variability of wastes make measurement of organic material or biomass costly, difficult, and uncertain. Even in solid-substrate fermentation, where frequently one microorganism is used, the theoretical approach is quite complicated to develop coupled with the heat and mass balance (Mitchell et al., 2004). Nevertheless, it has been shown that this approach can also be used for a process as complicated as biodrying.

Regarding the inductive strategy, to determine waste biodegradability and generate a useful measure for the loss of organic matter during biodrying, first-order kinetics has been proposed (Villegas and Huiliñir, 2014):

$$-r_{\text{BVS}} = \frac{d(\text{BVS})}{dt} = -k_T \cdot \text{BVS} \qquad (3.20)$$

The constant k_T is defined as the product of k (corrected first-order rate constant) multiplied by given environmental correction factors (Baptista et al., 2010; Haug, 1993):

$$k_T = k \cdot f(T) \cdot f(\text{MC}) \cdot f(O_2) \cdot f(\text{FAS}) \qquad (3.21)$$

where $f(T)$, $f(\text{MC})$, $f(O_2)$, and $f(\text{FAS})$ are the correction factors for temperature (T), moisture content (MC), oxygen concentration in the free air space (O_2), and free air space (FAS), respectively.

The correction factor for temperature, $f(T)$, has been described for an extensive set of equations in the literature on composting (Baptista et al., 2010). Expansions of the Arrhenius function, empirical functions, or functions based on cardinal (maximum, minimum, and optimum) temperatures are the more common (Mason, 2006). According to Richard and Walker (2006), the cardinal temperatures function presented by Rosso et al. (1993) provided the best description of the rate coefficient temperature dependence, since it involved the fewest parameters, all of which were easily measurable and each of which had a physical meaning in terms of the composting process (Mason, 2006). This function has also been used by Baptista et al. (2010) and Bari and Koenig (2012). The expansion of the Arrhenius function, proposed originally by Haug (1993), has also been widely used in models of sewage sludge composting (de Guardia et al., 2008; Kabbashi, 2011; Zhang et al., 2010), poultry manure (Petric and Selimbašić, 2008), and food waste (Richard and Walker, 2006). Nevertheless, for the biodrying process only the original function proposed by Haug (1993) has been used (Villegas and Huiliñir, 2014):

$$f(T) = \left(1,066^{(T-20)} - 1.21^{(T-60)}\right) \qquad (3.22)$$

The moisture content (MC) of the waste matrix is the single most important variable not only for evaluating the performance of biodrying, but also for controlling the optimal initial conditions of the waste matrix (Velis et al., 2009). Published moisture correction functions have all been derived empirically (Mason, 2006). The most popular function is the logistic function, proposed by Haug (1993), which is based on the trend of several sets of data from other authors for simulated food waste, municipal solid waste, and ground garbage composting. This function has also been used by several composting models (Baptista et al., 2010; Bari and Koenig, 2012; Petric and Selimbašić, 2008) and for the biodrying process (Huiliñir et al., 2017).

The effect of oxygen concentration has been modelled using Monod-type and exponential expressions (Haug, 1993; Richard et al., 2006). A simple one-parameter model was used by Haug (1993), with a half-saturation constant value of 2%. However, Richard et al. (2006) compared the performance of this model with two modified functions (modified one-parameter and two-parameter Monod models), showing that a modified one-parameter is justified for systems that frequently operate at near-ambient oxygen concentrations, particularly under low temperature and high moisture conditions. These characteristics are typical in biodrying systems, and therefore this modified one-parameter Monod model could be used in biodrying models. Until now, no proposed model had included the correction factor for oxygen, mainly because the oxygen under biodrying conditions (high airflow rate) is enough for the microorganisms' development.

Free air space (FAS) is important in biodrying because it is highly correlated with oxygen transfer in aerobic reactors. Data suggest that the optimum moisture content for a biological process is between 55 and 65%. That optimum exists because of rate-limiting moisture films which reduce oxygen diffusion in pilot scale and commercial scale systems (Ruggieri et al., 2009). Reaction rates proceed most rapidly in an aqueous environment of 100% moisture (used in the respirometric evaluation), but above 65%, pilot and commercial scale systems tend to become oxygen-limited, so the aerobic assumption is compromised. Several composting models (Baptista et al., 2010; Petric and Selimbašić, 2008; Zhang et al., 2010) include the FAS correction function based on the work of Haug (1993), who presented an inductive model based on the observation of the existence of a limited FAS range (>30%) that could sustain high activities (expressed as OUR) in the optimum moisture range. This function was also tested by Villegas and Huiliñir (2014) for the biodrying process.

Until now, only Huiliñir and Villegas (2014) and Villegas and Huiliñir (2014) had studied the kinetics of BVS degradation in a biodrying process using the inductive strategy and proposing new functions for the process. In the case of sewage sludge from paper and pulp industry wastewater treatment, Huiliñir and Villegas (2014) proposed a function that incorporated the factor corrections for T and MC, which successfully simulate the VS biodegradation under biodrying conditions, with a root mean square error (RMSE) between 0.007929 and 0.02744. The proposed function was:

$$\frac{d(\text{BVS})}{dt} = k_T \cdot \text{BVS} = a \cdot \left(b^{c \cdot T + d \cdot \frac{\text{MC}}{T}} \right) \tag{3.23}$$

where *a*, *b*, *c*, and *d* are parameters that should be found by fitting the function to experimental data.

Later, Villegas and Huiliñir (2015) presented a similar function applied to the biodrying of sewage sludge coming from a slaughterhouse wastewater treatment plant. This work improved the earlier one by including also the effect of FAS on the kinetics. Therefore, the new function proposed for several conditions of initial moisture content and airflow rate was

$$\frac{d(\mathrm{BVS})}{dt} = k_T \cdot \mathrm{BVS} = a \cdot \frac{\left(b^{c\cdot T + d\cdot \frac{\mathrm{MC}}{T}}\right)}{\exp(f\cdot \mathrm{FAS})} \cdot \mathrm{BVS} \tag{3.24}$$

Although the inductive strategy is easier to implement and has been widely explored, the deductive approach, as a mechanistic model, reflects the structure of the process, and is expected to yield better extrapolations (Hamelers, 2004). Furthermore, a deductive model can provide more flexibility in order to represent situations that could not be reproduced for inductive models (Villegas and Huiliñir, 2014). In recent years, Bialobrzewski et al. (2015), Petric and Mustafic (2015), Pujol et al. (2011), and He et al. (2018) used the deductive approach for modelling the composting process with good VS predictions and oxygen consumptions. In the case of biodrying, Huiliñir et al. (2017) proposed a kinetic model based on the model presented by Pommier et al. (2008) using the deductive approach. In this kinetic model the microbial population (X_H) was taken explicitly into account mainly through Monod-type kinetics, and total volatile solids (VS, measured in the laboratory) were divided into four different fractions: rapidly biodegradable (X_{rb}), slowly biodegradable (X_{sb}), non-biodegradable (or inert, X_I), and heterotrophic biomass (X_H). The X_{rb} and X_{sb} are hydrolyzed to the soluble substrate (S_S), which is consumed by the heterotrophic biomass. Therefore, S_S consumption generates new biomass. The biomass decay is also taken into account, and it generates rapidly biodegradable new VS. The VS degradation (X_{rb} and X_{sb} consumption) is considered an exothermic reaction, and it is the reaction that provides the energy used for drying. The kinetic model also included the effect of moisture content (MC) and temperature (T) in the process. A diagram of the conceptual model is presented in Figure 3.5.

Thus, up to date few reports have dealt with the kinetics of VS biodegradation, both the inductive and the deductive strategies. At this point it is still necessary to develop new research to improve the knowledge of this important topic.

3.5 RECOMMENDATIONS FOR FURTHER R&D

Even though in the last decade the information available about biodrying has greatly increased, there are still opportunities to improve not only the process itself, but also the knowledge of biological aspects and transport phenomena inside the matrix. One aspect that must be improved is the growth of biomass inside the bioreactor and how it is affected by operational conditions. Several efforts have been made in order to improve our knowledge of the different microorganisms inside these reactors; however, few attempts have been made to describe this mathematically. On the other hand, more efforts in modelling are necessary in order to establish new ways of controlling

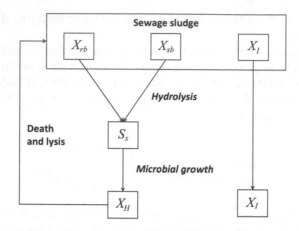

FIGURE 3.5 Conceptual kinetic model presented by Huiliñir et al. (2017).

the process. In this sense, a clear and compact kinetic model of the biological process that could relate energy release, VS degradation, and oxygen consumption is still necessary to achieve this task. Furthermore, techniques or different strategies to control the process in order to obtain better performance and efficiency also are required. Finally, studies about implementation at an industrial scale could also be important in order to improve the knowledge and limitations of this technology.

3.6 CONCLUSIONS

Biodrying is a biological process whose main advantage is the removal of water using the energy produced by an aerobic bioreaction, without external energy. Its application to sewage sludge drying had improved the management of sludge, including its recovery as a solid fuel. Because of these advantages, the scientific community has focused its interest on this process over the last two decades, with several results. One aspect of these results is related to the operational conditions; according to the literature the main parameters that must be considered for moisture removal are the amount of bulking agent, inlet airflow rate, and sludge moisture content. The bulking agent improves the porosity of the biomass matrix but decreases biological activity and bioheat. Therefore, the reduction of the rate of biological activity must be balanced against the increased porosity that the bulking agent favours. Airflow is necessary to produce heat through biological reactions and to remove water from the matrix, whose temperature affects the airflow temperature and subsequently its water holding capacity. Moisture content is related to the biological activity, with contents between 65 and 50% for an efficient biological activity.

Other aspects that have been studied are the energy and mass balances, in order to define how the energy coming from the biochemical reaction is used and how the water is distributed in the different streams of the process. Related to this aspect is the modelling of the process, where advances have also been reported, allowing an understanding of how the water is removed from the matrix and how the temperature, biodegradable solids, and moisture content are related to each other.

REFERENCES

Adani, F., Baido, D., Calcaterra, E., Genevini, P. 2002. "The influence of biomass temperature on biostabilization-biodrying of municipal solid waste". *Bioresource Technology*, 83(3), 173–179.

Adhikari, B.K., Barrington, S., Martinez, J., King, S. 2008. "Characterization of food waste and bulking agents for composting". *Waste Management*, 28(5), 795–804.

Adhikari, B.K., Barrington, S., Martinez, J., King, S. 2009. "Effectiveness of three bulking agents for food waste composting". *Waste Management*, 29(1), 197–203.

Agamuthu, P., Choong, L.C., Hasan, S., Praven, V.V. 2000. "Kinetic evaluation of composting of agricultural wastes". *Environmental Technology*, 21(2), 185–192.

Baader, W., Schuchardt, R., Sonnenberg, H. 1975. "Investigation in the development of technical management systems for the production of solids from animal waste". *Grundlagen der Landtechnik*, 25(2), 33–42.

Baptista, M., Antunes, F., Goncalves, M.S., Morvan, B., Silveira, A. 2010. "Composting kinetics in full-scale mechanical-biological treatment plants". *Waste Management*, 30(10), 1908–1921.

Bari, Q.H., Koenig, A. 2012. "Application of a simplified mathematical model to estimate the effect of forced aeration on composting in a closed system". *Waste Management*, 32(11), 2037–2045.

Barrington, S., Choiniere, D., Trigui, M., Knight, W. 2002. "Effect of carbon source on compost nitrogen and carbon losses". *Bioresource Technology*, 83(3), 189–194.

Bialobrzewski, I., Miks-Krajnik, M., Dach, J., Markowski, M., Czekala, W., Gluchowska, K. 2015. "Model of the sewage sludge-straw composting process integrating different heat generation capacities of mesophilic and thermophilic microorganisms". *Waste Management*, 43, 72–83.

Chang, J.I., Chen, Y.J. 2010. "Effects of bulking agents on food waste composting". *Bioresource Technology*, 101(15), 5917–5924.

Chen, G., Yue, P.L., Mujumdar, A.S. 2002. "Sludge dewatering and drying". *Drying Technology*, 20(4–5), 883–916.

Choi, H.L., Richard, T.L., Ahn, H.K. 2001. "Composting high moisture materials: Biodrying poultry manure in a sequentially fed reactor". *Compost Science & Utilization*, 9(4), 303–311.

Das, K.C., Tollner, E.W., Eiteman, M.A. 2003. "Comparison of synthetic and natural bulking agents in food waste composting". *Compost Science & Utilization*, 11(1), 27–35.

De Guardia, A., Petiot, C., Benoist, J.C., Druilhe, C. 2012. "Characterization and modelling of the heat transfers in a pilot-scale reactor during composting under forced aeration". *Waste Management*, 32(6), 1091–1105.

De Guardia, A., Petiot, C., Rogeau, D. 2008. "Influence of aeration rate and biodegradability fractionation on composting kinetics". *Waste Management*, 28(1), 73–84.

Figueroa-Montero, A., Esparza-Isunza, T., Saucedo-Castaneda, G., Huerta-Ochoa, S., Gutierrez-Rojas, M., Favela-Torres, E. 2011. "Improvement of heat removal in solid-state fermentation tray bioreactors by forced air convection". *Journal of Chemical Technology and Biotechnology*, 86(10), 1321–1331.

Frei, K.M., Cameron, D., Stuart, P.R. 2004. "Novel drying process using forced aeration through a porous biomass matrix". *Drying Technology*, 22(5), 1191–1215.

Gea, T., Barrena, R., Artola, A., Sanchez, A. 2007. "Optimal bulking agent particle size and usage for heat retention and disinfection in domestic wastewater sludge composting". *Waste Management*, 27(9), 1108–1116.

Hamelers, H.V.M. 2004. "Modeling composting kinetics: A review of approaches". *Reviews in Environmental Science and Biotechnology*, 3(4), 331–342.

Haug, R.T. 1993. *The Practical Handbook of Compost Engineering*. 2 ed. CRC Press LLC, Florida.

He, P.J., Zhao, L., Zheng, W., Wu, D., Shao, L.M. 2013. "Energy balance of a biodrying process for organic wastes of high moisture content: A review". *Drying Technology*, 31(2), 132–145.

He, X.Q., Han, L.J., Ge, J.Y., Huang, G.Q. 2018. "Modelling for reactor-style aerobic composting based on coupling theory of mass-heat-momentum transport and Contois equation". *Bioresource Technology*, 253, 165–174.

Huiliñir, C., Perez, J., Olivares, D. 2017. "A new model of batch biodrying of sewage sludge, Part 1: Model development and simulations". *Drying Technology*, 35(6), 651–665.

Huiliñir, C., Villegas, M. 2014. "Biodrying of pulp and paper secondary sludge: Kinetics of volatile solids biodegradation". *Bioresource Technology*, 157, 206–213.

Huiliñir, C., Villegas, M. 2015. "Simultaneous effect of initial moisture content and airflow rate on biodrying of sewage sludge". *Water Research*, 82, 118–128.

Jewell, W.J., Donder, N.C., van Soest, P.J., Cummings, R.T., Vergara, W.W., Linkenheil, R. 1984. "High temperature stabilization and moisture removal from animal wastes for by-product recovery". Cooperative State Research Service, SEA/CR 616-15-168.

Kabbashi, N. 2011. "Sewage sludge composting simulation as carbon/nitrogen concentration change". *Journal of Environmental Sciences-China*, 23(11), 1925–1928.

Kim, K.Y., Kim, H.W., Han, S.K., Hwang, E.J., Lee, C.Y., Shin, H.S. 2008. "Effect of granular porous media on the composting of swine manure". *Waste Management*, 28(11), 2336–2343.

Kulcu, R., Yaldiz, O. 2007. "Composting of goat manure and wheat straw using pine cones as a bulking agent". *Bioresource Technology*, 98(14), 2700–2704.

Larsen, K.L., McCartney, D.M. 2000. "Effect of C: N ratio on microbial activity and N retention: Bench-scale study using pulp and paper biosolids". *Compost Science & Utilization*, 8(2), 147–159.

Li, X.W., Dai, X.H., Yuan, S.J., Li, N., Liu, Z.G., Jin, J.W. 2015. "Thermal analysis and 454 pyrosequencing to evaluate the performance and mechanisms for deep stabilization and reduction of high-solid anaerobically digested sludge using biodrying process". *Bioresource Technology*, 175, 245–253.

Liang, C., Das, K.C., McClendon, R.W. 2003. "The influence of temperature and moisture contents regimes on the aerobic microbial activity of a biosolids composting blend". *Bioresource Technology*, 86(2), 131–137.

Lin, Y.P., Huang, G.H., Lu, H.W., He, L. 2008. "Modeling of substrate degradation and oxygen consumption in waste composting processes". *Waste Management*, 28(8), 1375–1385.

Ma, J., Zhang, L., Li, A.M. 2016. "Energy-efficient co-biodrying of dewatered sludge and food waste: Synergistic enhancement and variables investigation". *Waste Management*, 56, 411–422.

Mason, I.G. 2006. "Mathematical modelling of the composting process: A review". *Waste Management*, 26(1), 3–21.

Miller, F.C. 1989. "Matric water potential as an ecological determinant in compost, a substrate dense system". *Microbial Ecology*, 18(1), 59–71.

Mitchell, D.A., von Meien, O.F., Krieger, N., Dalsenter, F.D.H. 2004. "A review of recent developments in modeling of microbial growth kinetics and intraparticle phenomena in solid-state fermentation". *Biochemical Engineering Journal*, 17(1), 15–26.

Navaee-Ardeh, S., Bertrand, F., Stuart, P.R. 2006. "Emerging biodrying technology for the drying of pulp and paper mixed sludges". *Drying Technology*, 24(7), 863–878.

Navaee-Ardeh, S., Bertrand, F., Stuart, P.R. 2010. "Key variables analysis of a novel continuous biodrying process for drying mixed sludge". *Bioresource Technology*, 101(10), 3379–3387.

Navaee-Ardeh, S., Bertrand, F., Stuart, P.R. 2011a. "A 2D distributed model of transport phenomena in a porous media biodrying reactor". *Drying Technology*, 29(2), 153–162.

Navaee-Ardeh, S., Bertrand, F., Stuart, P.R. 2011b. "Development and experimental evaluation of a 1D distributed model of transport phenomena in a continuous biodrying process for pulp and paper mixed sludge". *Drying Technology*, 29(2), 135–152.

Navaee-Ardeh, S., Bertrand, F., Stuart, P.R. 2011c. "A 2D distributed model of transport phenomena in a porous media biodrying reactor". *Drying Technology: An International Journal*, 29(2), 153–162.

Ofori-Boateng, C., Lee, K.T., Mensah, M. 2013. "The prospects of electricity generation from municipal solid waste (MSW) in Ghana: A better waste management option". *Fuel Processing Technology*, 110, 94–102.

Pasda, N., Limtong, P., Oliver, R., Montange, D., Panichsakpatana, S. 2005. "Influence of bulking agents and microbial activator on thermophilic aerobic transformation of sewage sludge". *Environmental Technology*, 26(10), 1127–1135.

Petric, I., Mustafic, N. 2015. "Dynamic modeling the composting process of the mixture of poultry manure and wheat straw". *Journal of Environmental Management*, 161, 392–401.

Petric, I., Selimbašić, V. 2008. "Development and validation of mathematical model for aerobic composting process". *Chemical Engineering Journal*, 139(2), 304–317.

Pommier, S., Chenu, D., Quintard, M., Lefebvre, X. 2008. "Modelling of moisture-dependent aerobic degradation of solid waste". *Waste Management*, 28(7), 1188–1200.

Pujol, A., Debenest, G., Pommier, S., Quintard, M., Chenu, D. 2011. "Modeling composting processes with local equilibrium and local non-equilibrium approaches for water exchange terms". *Drying Technology*, 29(16), 1941–1953.

Rada, E.C., Franzinelli, A., Taiss, M., Ragazzi, M., Panaitescu, V., Apostol, T. 2007. "Lower heating value dynamics during municipal solid waste biodrying". *Environmental Technology*, 28(4), 463–469.

Rada, E.C., Ragazzi, M., Fiori, L., Antolini, D. 2009. "Biodrying of grape marc and other biomass: a comparison". *Water Science and Technology*, 60(4), 1065–1070.

Richard, T.L., Hamelers, H.V.M., Veeken, A., Silva, T. 2002. "Moisture relationships in composting processes". *Compost Science & Utilization*, 10(4), 286–302.

Richard, T.L., Veeken, A.H.M., de Wilde, V., Hamelers, H.V.M. 2004. "Air-filled porosity and permeability relationships during solid-state fermentation". *Biotechnology Progress*, 20(5), 1372–1381.

Richard, T.L., Walker, L.P. 2006. "Modeling the temperature kinetics of aerobic solid-state biodegradation". *Biotechnology Progress*, 22(1), 70–77.

Richard, T.L., Walker, L.P., Gossett, J.M. 2006. "Effects of oxygen on aerobic solid-state biodegradation kinetics". *Biotechnology Progress*, 22(1), 60–69.

Rosso, L., Lobry, J.R., Flandrois, J.P. 1993. "An unexpected correlation between cardinal temperatures of microbial growth highlighted by a new model". *Journal of Theoretical Biology*, 162(4), 447–463.

Ruggieri, L., Gea, T., Artola, A., Sanchez, A. 2009. "Air filled porosity measurements by air pycnometry in the composting process: A review and a correlation analysis". *Bioresource Technology*, 100(10), 2655–2666.

Rulkens, W. 2008. "Sewage sludge as a biomass resource for the production of energy: Overview and assessment of the various options". *Energy & Fuels*, 22(1), 9–15.

Stasta, P., Boran, J., Bebar, L., Stehlik, P., Oral, J. 2006. "Thermal processing of sewage sludge". *Applied Thermal Engineering*, 26, 1420–1426.

Sugni, M., Calcaterra, E., Adani, F. 2005. "Biostabilization-biodrying of municipal solid waste by inverting air-flow". *Bioresource Technology*, 96(12), 1331–1337.

Tremier, A., de Guardia, A., Massiani, C., Paul, E., Martel, J.L. 2005. "A respirometric method for characterising the organic composition and biodegradation kinetics and the temperature influence on the biodegradation kinetics, for a mixture of sludge and bulking agent to be co-composted". *Bioresource Technology*, 96(2), 169–180.

Tremier, A., Teglia, C., Barrington, S. 2009. "Effect of initial physical characteristics on sludge compost performance". *Bioresource Technology*, 100(15), 3751–3758.

Treybal, R. 1986. *Mass Transfer Operations*. 3 ed. McGraw-Hill, Mexico.

Velis, C.A., Longhurst, P.J., Drew, G.H., Smith, R., Pollard, S.J.T. 2009. "Biodrying for mechanical-biological treatment of wastes: A review of process science and engineering". *Bioresource Technology*, 100(11), 2747–2761.

Villegas, M., Huiliñir, C. 2014. "Biodrying of sewage sludge: Kinetics of volatile solids degradation under different initial moisture contents and air-flow rates". *Bioresource Technology*, 174, 33–41.

Winkler, M.K.H., Bennenbroek, M.H., Horstink, F.H., van Loosdrecht, M.C.M., van de Pol, G.J. 2013. "The biodrying concept: An innovative technology creating energy from sewage sludge". *Bioresource Technology*, 147, 124–129.

Yang, B., Zhang, L., Jahng, D. 2014. "Importance of initial moisture content and bulking agent for biodrying sewage sludge". *Drying Technology*, 32(2), 135–144.

Yang, B.Q., Hao, Z.D., Jahng, D. 2017. "Advances in biodrying technologies for converting organic wastes into solid fuel". *Drying Technology*, 35(16), 1950–1969.

Zambra, C.E., Rosales, C., Moraga, N.O., Ragazzi, M. 2011. "Self-heating in a bioreactor: Coupling of heat and mass transfer with turbulent convection". *International Journal of Heat and Mass Transfer*, 54(23–24), 5077–5086.

Zavala, M.A.L., Funamizu, N., Takakuwa, T. 2004. "Modeling of aerobic biodegradation of feces using sawdust as a matrix". *Water Research*, 38(5), 1327–1339.

Zawadzka, A., Krzystek, L., Stolarek, P., Ledakowicz, S. 2010. "Biodrying of organic fraction of municipal solid wastes". *Drying Technology*, 28(10), 1220–1226.

Zhang, J., Gao, D., Chen, T.B., Zheng, G.D., Chen, J., Ma, C.A., Guo, S.L., Du, W. 2010. "Simulation of substrate degradation in composting of sewage sludge". *Waste Management*, 30(10), 1931–1938.

Zhang, Y., Lashermes, G., Houot, S., Doublet, J., Steyer, J.P., Zhu, Y.G., Barriuso, E., Garnier, P. 2012. "Modelling of organic matter dynamics during the composting process". *Waste Management*, 32(1), 19–30.

Zhao, L., Gu, W.-M., He, P.-J., Shao, L.-M. 2010. "Effect of air-flow rate and turning frequency on biodrying of dewatered sludge". *Water Research*, 44(20), 6144–6152.

Zhao, L., Gu, W.-M., He, P.-J., Shao, L.-M. 2011. "Biodegradation potential of bulking agents used in sludge biodrying and their contribution to bio-generated heat". *Water Research*, 45(6), 2322–2330.

4 Advances in Algae Dewatering Technologies

K.Y. Show, Y.G. Yan, and D.J. Lee

CONTENTS

4.1 INTRODUCTION

Biofuels are perceived as favorable replacements for fossil fuels in alleviating environmental problems such as carbon emissions. Among the crops, algae are considered more promising because of their effective transformation of sunlight into chemical energy (Deirue et al., 2012; Lee et al., 2012; Liu et al., 2012; Show and Lee, 2013). Algal biomass can produce much higher biofuel yield per crop area, and it does not compete with other food crops such as soybean and corn. As CO_2 captured in algae is up to two orders of degree more in than terrestrial vegetation (Wang et al., 2008), algae stand an important position in the carbon cycle by estranging surplus CO_2 from the atmosphere.

Processing biodiesel from algae entails a series of operations involving algal cultivation, harvesting, concentration, drying, oil extraction, and fractionation (Zheng et al., 2012; Show et al., 2013). This chapter provides a review and deliberation on the latest development in algal harvesting through separation and concentration. Since algae screening and thickening are the primary processes of concentration (Figure 4.1), the discussion on concentration involves screening, thickening, and dewatering. Challenges and prospects with regard to algae harvesting are also outlined. The main purpose of this review is to provide pertinent information for further development of practical production of algal biofuel.

4.2 STABILITY AND SEPARATION OF ALGAE

Harvesting of algae is basically a process entailing the separation of the biomass from the growth media followed by concentration. Primary factors affecting the stability of microalgae are closely associated with surface charge, size, and density of the algal cells. And the stability in turn dictates algae separation and concentration from the growth medium. It has been reported that broth electro-interactions among algal cells and cell interactions with the ambient influence the surface charge of the algal in suspension (Tenney et al., 1969). In contrast, algae size and density dominate the settling velocity of the cells, and this is an important consideration in process selection for algae separation through thickening.

The influence of ozonation on surface charge and particle stability was examined (Farvardin and Collins, 1989; Chheda and Grasso, 1994). Theories for improvement on coagulation of suspended solids like algae due to ozonation were hypothesized (Reckhow et al., 1986; Plummer and Edzwald, 2002), and it was deduced that applying suitable ozonation could lead to destabilization of particles which in turn enhanced separation from the suspension.

Algogenic organic matters (AOM) derived from four species of algae, namely *Asterionella formosa, Chlorella vulgaris, Melosira* sp., and *Microcystis aeruginosa*, were dominated by hydrophilic polysaccharides and hydrophobic proteins of negative zeta potentials and low specific ultraviolet absorbance (Henderson et al., 2008). AOM hydrophobicity was mainly owing to the hydrophobic proteins with molecular mass heavier than 500 kDa. The charge density, however, decreased inversely with hydrophobicity. The authors explained that the charge density was associated with hydrophilic and acidic carbohydrates rather than hydrophobic humic acids. Subsequently, it was found that treated bubbles could separate algae satisfactorily

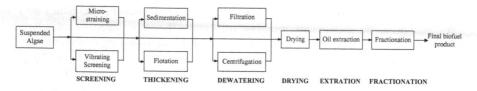

FIGURE 4.1 Schematic flow processes for algal biofuel production.

without prior coagulation and flocculation (Henderson et al., 2009). The surfaces of the bubbles were treated with chemicals of both a hydrophilic high charge head and a hydrophobic long tail.

Organic polymers were able to destabilize colloidal algal suspension from specific ionic or atomic interactions (Shelef et al., 1984). Long-chain industrial polymers such as polyelectrolytes or polyhydroxy compounds are able to better perform coagulation or flocculation. Such polymeric coagulation-flocculation was associated with a bridging model hypothesizing that a polymer clings onto algal surfaces via some segments while the remaining segments protrude into the solution (Gregory, 1977). These segments are then able to bridge on vacant sites of other algal cells, forming a three-dimensional aggregate network that enhances algae settling and separation.

Clearly, a larger aggregation body of algae cells can increase the overall biomass settling rate. Based on this principle, chemical coagulants are applied in algal separation to form large and heavy flocs which settle rapidly by gravity to the bottom of the tank. A similar principle can be illustrated by applying centrifugation on algal suspensions to separate algal cells. Further deliberation on this notion will be discussed in the subsequent section on coagulation-flocculation.

In summary, destabilization of algal suspension is an important pretreatment for most of the algal separation and concentration processes described in Section 4.3.

4.3 ALGAE CONCENTRATION PROCESSES

Separating algae from the culture medium and subsequent concentration is known as harvesting. For microalgae grown in aqueous medium, concentrating loose algae grown in suspension to a dense slurry or cake is an important process of harvesting. The water held within algae must be removed as far as possible for practical harvesting and downstream dewatering, drying, oil extraction, and fractionation (Figure 4.1). Development of the concentration process alternatives is reviewed and discussed in the following sections.

4.3.1 SCREENING

The first treatment process of harvesting is screening whereby algal biomass is passed through a screen of particular openings. Screening performance and output algae concentration are dictated by the sizes of algae and screen aperture. Adequate design of the screen produces superior algae separation and denser algae output. Common screening methods include vibrating screens and microstrainers.

4.3.2 VIBRATING SCREEN

Earlier work on algae strains *Coelastrum* harvesting by vibrating screen was carried out by Mohn (1980). Denser algae concentration of 7–8% has been reported under batch process compared with thinner contents of 5–6% harvested by continuous mode. In a commercial application, vibrating screens were employed for harvesting *Spirulina* as a food supply (Habib et al., 2008). The commercial *Spirulina* belong to two distinct genera: *Spirulina* and *Arthrospira*, which are multicellular and filamentous blue-green microalgae. The vibrating screen achieved up to 95% harvesting producing algal solid contents of 8–10% with a harvesting rate of 20 m³/h. The vibrating screens required a relatively smaller footprint occupying only one-third of the area of inclining screens with an area of 2–4 m²/unit.

4.3.3 MICROSTRAINING

Microstrainers are constructed with a rotary drum enclosed by a straining fabric made of either stainless steel or polyester. Algae of sizes smaller than the strainer could still pass through unharvested. The solids content of algae harvested through microstraining is still low and further dewatering is usually incorporated.

Some success in separating algae from pond effluent with non-stop backwashing in microstrainers was reported (Koopman et al., 1978; Shelef et al., 1980). The success was limited to larger size algae species such as *Micractinium* and *Scenedesmus*, as the smallest strainer openings at that time were confined to 23 μm. Subsequently, greater success in decreasing solids concentration in an algae lagoon effluent from about 80 to below 20 mg/L by rotary microstrainers with 1 μm screen was achieved (Wettman and Cravens, 1980). Concentrating *Coelastrum proboscideum* to about 1.5% solids content using microstrainers operating at a cost of about DM 0.02/m³ and an energy consumption of 0.2 kWh/m³ was reported (Mohn, 1980).

The shortfalls of microstrainers include low harvesting performance and problems in handling algae with variations in size. These problems can be solved partly by changing the rotation speed of the drum (Reynolds et al., 1975). Another difficulty of microstraining is biofilm build-up on the fabric or screen. Periodic clean-up of the fabric or mesh may prevent such biofilm build-up.

4.4 THICKENING

Thickening is a treatment following screening with the objective to raise the solids content of the algae for viable downstream processes such as dewatering. With even a moderate rise in algae solids content, thickening is economically advantageous as a significant reduction in biomass volume can be accrued. This reduction can bring about considerable cost savings in subsequent treatments. Regular methods for thickening can be generally grouped into sedimentation and flotation, as outlined in Figure 4.2.

4.4.1 SEDIMENTATION

Process options for algae sedimentation comprise gravity sedimentation, coagulation-flocculation, ultrasound sedimentation, and electro-flocculation. Gravity sedimentation exploits gravity force to achieve solid-liquid separation of algae suspension. The content of algae suspension is segregated into two streams: deposition of denser algal biomass at the bottom of the sedimentation tank and clarified effluent overflowing at the top. The use of gravity sedimentation can either be directed to achieve clear effluent with dilute feeds (Mohn, 1980; Sukenik and Shelf, 1984), or to achieve denser algae deposition with more concentrated feeds (Mohn, 1980).

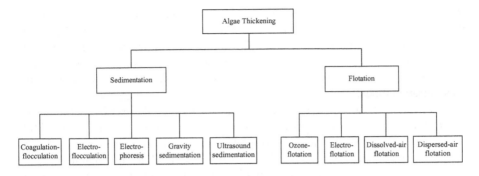

FIGURE 4.2 Process alternatives for algae thickening.

Reports on algae sedimentation in ponds without the aid of flocculation are limited. Processes adopting fill-and-draw series for the secondary pond resulted in the considerable separation of algae from facultative-oxidative pond effluent (Benemann et al., 1980). Identical secondary ponds were studied for algae sedimentation fed from high-rate oxidative pond effluents (Adan and Lee, 1980; Benemann et al., 1980). Satisfactory effluent quality and deposits concentrations (up to 3%) were derived from algae self-flocculation. Self-flocculation or natural flocculation of algae may be exploited as a cheap and simple separation technique. However, its performance and reliability were low without the aid of chemical coagulant. Batched sedimentation achieving an algal deposit of 1.5% solids content with coagulant aids was reported (Mohn, 1980).

Inclined flat plates were mounted on a settling basin to enhance solids aggregation and settling (Mohn, 1980; Shelef et al., 1984). The inclinations of the plates were designed to direct settling of algae along the slopes towards a sump. Settled algae were collected in the sump from which they were sucked out by a pump. Algae concentration of 1.6% solids content was achieved. For separation of tiny algae such as *Scenedesmus,* coagulant dosing was suggested to enhance sedimentation (Mohn, 1980). The process reliability of this technique was reasonable. Further thickening of the settled slurry was usually needed.

The effects of coagulation-flocculation pretreatment for gravity sedimentation have been investigated (Golueke and Oswald, 1965). Applying aluminium sulfate (commonly known as alum) as a coagulating agent in coagulation-flocculation, the sedimentation collected up to 85% of the algal suspension content in treating high-rate oxidative pond effluent. The flocculation-sedimentation process was reliable, and a variety of algal species had been treated to attain algae slurry of 1.5% solids concentration. On the other hand, an evaluation of the flocculation-sedimentation operation in comparison with the flocculation-flotation process suggested a superiority of the latter, demonstrating definite optimum process parameter for algae separation (Friedman et al., 1977; Moraine et al., 1980). Further delineation on flotation will be delivered in Section 4.4.2.

The use of acoustic techniques for algae separation and concentration has been examined (Bosma et al., 2003). Separation based upon ultrasonic-induced agglomeration of algal cells accompanied by sedimentation was conducted. Algae removal in excess of 90% was achieved with feed flows between 4 and 6 L/d. Up to 92% of the algal biosolids could be separated with a concentration index of 11. Efforts to improve the separation were unsuccessful due to the minute size and lightweight nature of the algal particles. Other factors such as feed flow, feed concentration, and proportion between harvest and feed flow rates exerted a major effect on the concentration index.

Application of sonication to improve separation by coagulation of a common toxic algal species, namely *Microcystis aeruginosa*, was examined (Zhang et al., 2009). The results demonstrated that ultrasound treatment promoted substantially the reduction of algal particles, aqueous UV_{254}, and chlorophyll *a* while maintaining the level of aqueous microcystins. A mechanism was proposed that destruction within algal cells induced by both acoustic cavitation and molecular collapse would result in the sedimentation of cyanobacteria. The study revealed that separation efficiency relied highly on the coagulant dosage and ultrasound operating conditions. Using a coagulant dosage of 0.5 mg/L and sonication for 5 sec, algae separation markedly enhanced from 35 to 67%. It was found that the optimal sonication duration was 5 sec; further sonication did not improve the performance further. Sonication at 47.2 W/cm^2 was found to be most effective for maximum algae removal of 93.5%. It was suggested that this method be employed on natural cultures containing a variety of algae species.

Besides a small footprint, experimental ultrasound studies showed that the integrity of the algae can be maintained, because the method can be operated continuously without causing hydrodynamic shear on algal cells (Bosma et al., 2003). The authors suggested the use of centrifuges over the ultrasonic sedimentation for commercial harvesting because of lower energy consumption, and better separation and concentration factors.

Coagulation-flocculation induces algal cells to chemically bind together forming larger aggregates that can be better filtered or settled. Algal coagulants can be generally categorized into two groups, namely organic and inorganic coagulants.

Inorganic cationic coagulants such as Al^{+3} and Fe^{+3} form polyhydroxy complexes at specific pHs. Aluminium sulfate (or alum) or other salts of aluminium are common coagulants used in water treatment and algae separation. Ferric sulfate is another coagulant commonly used; however, it was found to be inferior in comparison with alum in terms of optimal dosage, pH, and product quality (Bare et al., 1975; Moraine et al., 1980).

Satisfactory coagulation normally occurs at optimum pHs. Hydrated lime is commonly used to raise the pH to the optimum level at which an inorganic substance (magnesium hydroxide) is formed. As magnesium hydroxide reacts as a coagulant, the lime is functioning more like a coagulant inducer (Friedman et al., 1977). A satisfactory performance in algae harvesting with the addition of lime has been reported (Shelef et al., 1984; Friedman et al., 1977). Nevertheless, the performance was confined to feed containing Mg^{2+} above 10 mg/L, while the product quality was badly affected by excessive Ca^{2+} concentration (up to 25% by weight).

A well accepted hypothesis maintained that cationic coagulants neutralized the repelling negative surface charge of algal particles, driving them to coalesce into aggregates. It follows that more coagulant is required for better coagulation in a linear stoichiometric manner. This makes coagulation an expensive process. Conversely, a study disputed that the optimum quantity of cationic coagulant was not related to electrostatic charges. It was claimed that the optimum level was a function of the logarithm of particle density – alkaline requirement of dense algal cultures was an order of magnitude lower than dilute cultures with coagulation taking place at a lower pH (Schlesinger et al., 2012).

Mechanisms to explain algal coagulation-flocculation derived from the application of polymeric coagulants were proposed. Cationic polyelectrolytes of larger molecular mass performed better in flocculating algal cells than their lower molecular mass counterparts. Polymer dosing diminished with increasing the molecular mass of the polyelectrolyte, and high molecular mass polyelectrolytes may better neutralize the cellular surface charge, thus promoting sedimentation. The work also indicated that the polyelectrolyte dosage required depended on feed concentrations, and the relationship between dosage and feed concentration can be determined by stoichiometry.

A more expensive commercial coagulant, "chitosan," has long been used for water purification. Chitosan was also tested effectively in algae flocculation and settling (Divakaran and Pillai, 2002). It was reported that the flocculation performance was very susceptible to pH with a narrow optimal pH range. Optimum pH at 7.0 was found for superior flocculation of freshwater algae. Based on three freshwater algae (*Spirulina*, *Oscillatoria*, and *Chlorella*) and one brackish algae (*Synechocystis*) with suspensions varying in the pH range 4–9 and chlorophyll *a* contents ranging between 80 and 800 mg/m^3, the chitosan-assisted flocculation accomplished a clarified effluent turbidity of 10–100 NTU units.

A fungi pelletizing-aided biological flocculation method for algae harvesting and wastewater purification was formulated (Zhou et al., 2012). The main factor affecting the development of fungi-algae pellets was found to be pH, which could be regulated by altering concentrations of glucose and fungal spores. Greatest pelletization developed when dosing 20 g/L glucose and 1.2×10^8/L spores in BG-11 broth medium, whereby most algal particles were picked up by the pellets with lower reaction time. The larger size (2–5 mm) fungi-algae pellets can be readily harvested by plain filtration.

Harvesting through chemical flocculation is often an expensive process for large-scale production. Another problem of this harvesting method is the difficulty in removing excess chemicals from the harvested algae, which makes it wasteful and unprofitable for commercial application. A way to solve this setback was to truncate the CO_2 supply to the algal photosynthetic system causing "auto-flocculation" of the algae. In certain circumstances, auto-flocculation was linked to elevated pH due to photosynthetic CO_2 intake leading to the formation of inorganic precipitates (predominantly calcium phosphate) that induced flocculation (Sukenik and Shelef, 1984).

There are numerous parameters that could collectively influence algal coagulation-flocculation in a complex manner. Hence, prediction for the operating conditions is nearly impracticable. Besides types of algae, optimal coagulant quantities could also be affected by levels of alkalinity, phosphate, dissolved organic matter, ammonia, and temperature of the algal suspension (Moraine et al., 1980; Shelef et al., 1980, 1984). In practice, optimal coagulant doses are best ascertained through bench-scale jar tests, which are devised to simulate multifaceted coagulation-floc-culation processes.

Electrical methods such as electro-flocculation and electrophoresis have been used to harvest algae. In an aquatic suspension, however, both electro-flocculation and electrophoresis could be operated under similar sets of conditions. If a basin of algae in suspension was subject to an electric flow by immersing metallic elec-trodes on both sides of the basin energized with a dc voltage, algae agglomerations would take place at both electrodes via electrophoresis and at the bottom of the basin through electro-flocculation.

Parameters affecting electrophoresis and electro-flocculation of algae in suspen-sion were assessed (Pearsall et al., 2011). The assessment revealed that electropho-resis was complicated by the fluid motion, and the connection between algal cell and fluid could be adequately intense such that the fluid motion could dominate the electrophoresis. While electro-flocculation seemed to be robust (Azarian et al., 2007), it does intrinsically accumulate electrically produced trace metal flocculants in the separated algae.

4.4.2 FLOTATION

Flotation encompasses different methods such as dispersed-air flotation, dissolved-air flotation, ozone flotation, and electroflotation. While gravity sedimentation works well with dense algae, flotation is suitable for separation of fluffy algal cells (Chen et al., 1998). These light and tiny cells settle so slowly that separating them by gravity sedimentation is impractical. Flotation is merely gravity concentration upside-down. Instead of letting the algal cells settle downwards, algal separation was accomplished by injecting air bubbles underneath the flotation vessel. The algal par-ticles attach themselves to the bubbles, and the collective buoyancy force promotes the flotation of algal cells to the liquid's surface. Formation of concentrated slurry takes place above the vessel, and is subsequently collected by a skimmer.

Coagulant can be used to enhance algae separation by flotation (Bare et al., 1975). Different chemicals such as ferric and aluminium salts and polymers were employed to assist flotation, hoping to increase solids loading rates, concentration of floated algae, and effluent clarity.

The major merit of flotation over sedimentation is that minute or fluffy algal cells that settle slowly by gravity can be harvested much faster. Furthermore, flotation also renders higher solids contents and lower equipment investment. There are four vari-ances of flotation processes, namely dissolved-air flotation, ozone flotation, electro-flotation (also termed electrolytic flotation), and dispersed-air flotation.

In the dissolved-air flotation (DAF) process, a pressurized liquid is pumped to the DAF tank where it is mixed with algal feed. As the pressure reverses to atmo-spheric level, the pressurized air emerges from the mixed solution generating fine bubbles. Fine algal particles attach onto the bubbles, and float to the top where they are skimmed and harvested.

Application of DAF method in combination with chemical flocculation has been reported for algae separation (McGarry and Durrani, 1970; Bare et al., 1975; Sandbank, 1979). The overflow effluent quality relied on process parameters such as recirculation ratio, air pressure, liquid residence time, and solids floating rate (Bare et al., 1975; Sandbank, 1979). On the other hand, algal slurry thickness was dictated by skimming rate and skimmer overboard on top of water surface (Moraine et al., 1980).

An algal suspension containing a variety of algal species could be concentrated by DAF attaining harvested slurry up to 6%. The solids content of slurry could be further raised by a subsequent second stage flotation (Bare et al., 1975; Friedman et al., 1977; Moraine et al., 1980). Optimal operating factors have to be established for achieving good reliability of DAF separation.

Dispersed-air flotation is a variant of DAF in which non-pressurized air is directly supplied to the flotation vessel. Large bubbles are created by a combination of agitation with air injection (froth-flotation) or by pumping air through perforated media (foam-flotation). In the froth-flotation process, the cultivator supplies aerated water to the froth, and concentrated algae is skimmed at the top. The performance of algae harvesting using froth-flotation is dictated by the type of collector, rates of aeration, pH, and temperature of algal medium (Phoochinda and White, 2003; Phoochinda et al., 2005). Supplement of surfactants such as cetyltrimethylammonium bromide (CTAB) and sodium dodecylsulfate (SDS) improved the aeration and bubble size. The CTAB showed high harvesting efficiency (90%) but SDS indicated poor performance at only 16%. The efficiency, however, was improved (by up to 80%) by lowering the pH of the algal medium. In another study, it was reported that only two out of 18 tested reagents yielded satisfactory algae concentration (Golueke and Oswald, 1965). The harvesting performance was mainly dictated by suspension pH in the dispersed-air flotation operation (Levin et al., 1962). Crucial pH value was determined at 4.0 due to the changes in the cellular surface characteristics.

Exploitation of ozone-induced flotation for algae harvesting and effluent purification was examined (Betzer et al., 1980; Benoufella et al., 1994; Jin et al., 2006; Cheng et al., 2010, 2011). The principle of ozonation entails particle flotation by altering the cellular wall surface and liberating certain surface-active substances from algal cells.

The removal of cyanobacteria (also called blue-green algae) *Microcystis* strain, characteristics of ozone oxidizing, and physical features of ozone flotation were investigated in a pilot trial (Benoufella et al., 1994). The trial work found that ozonation pretreatment resulted in suppression of algal cells. Upstream coagulation was needed for acceptable *Microcystis* removal, with coagulant agent ferric chloride giving the best results. Prior ozonation also affected the coagulation performance. Coupling the ozone flotation with subsequent filtration could enhance treated effluent quality with high clarity and low organic substance.

An algal species, *Scenedesmus obliquus* FSP-3, known for its great potentials for CO_2 sequestration and lipid generation, was harvested by dispersed-ozone flotation. The optimum ozone dosage used was similar to that applied in drinking water treatment. Dispersed-ozone flotation generates more efficient algae separation than its dispersed-air counterpart (Cheng et al., 2011). The effects of ozonation on algae removal, fluorescence properties, cellular surface charge and hydrophobicity, polysaccharides, and protein concentrations in algogenic organic matter (AOM) were examined. The study revealed that proteins liberated from securely bound AOM were necessary for the modification of the hydrophobicity of bubble surfaces. The modification brought about enhanced particle attachment and formation of a top froth layer for the collection of floating cells. The ozone introduced, however, was stripped by humic compounds present in the suspension. Ozone flotation efficiency was thus adversely affected.

In electroflotation (also called electrolytic flotation), algal cells attach onto H_2 bubbles generated during the electrolysis and float up to the liquid surface where they are harvested by a skimmer. Algae flocculation derived from magnesium hydroxide formed in bench-scale electroflotation electrolysis was reported (Contreras et al., 1981). Algae separation from effluent of the wastewater oxidation pond was examined using both laboratory-scale and field electroflotation systems (Sandbank et al., 1974; Kumar et al., 1981). In another work, a 2 m² pilot-unit was tested for effluent purification of a high-rate oxidation pond (Shelef et al., 1984). It was found that, for adequate algae separation, coupling electroflotation with alum flocculation is a good option (Sandbank et al., 1974).

A variety of algal species were harvested by electroflotation, achieving up to 5% solids contents in the harvested product. Subsequent decantation further raised the solids content to 7–8% (Sandbank, 1979). An expensive rectifier delivering 5–20 DC volts at approximately 11 A/m² to electrodes in place of a saturator was tested. The voltage needed to sustain the current density for the production of bubbles was dictated by the conductivity of the algal feed. Energy consumption of electroflotation is normally high, but the operating cost is economical for small setups (<5 m² area) compared to the DAF unit (Svarovsky, 1979).

4.5 DEWATERING

Thickening is considered a concentration process to increase the solids content of loose algae (usually up to 10% solids) to twofold or more. Dewatering has the same function of stepping up the solids concentration but is designed to remove water from algae in the highest quantity. Loose algae, say of 1–3% dry solids, could be thickened to about 2–6% dry solids. Such thickened algae could be further dewatered to 12–35% dry solids, at which dewatered state the harvested algae appear in semi-solid cake form rather than flowing liquid, and this can be easily handled manually or by mechanical means. Several mechanical machines devising different process alternatives of filtration and centrifugation (Figure 4.3) are used for dewatering.

4.5.1 FILTRATION

Filtration is implemented by pumping algal feed to flow across filter mesh. Algal biomass is trapped and concentrated on the mesh which is then harvested. The main merit of filtration is that very loose algal cells or microalgae can be harvested with satisfactory results. A pressure gradient must be kept in pumping fluid across the mesh. Various filtration processes including gravity, vacuum, pressure, or magnetic filtration with respectively designed pressure drops have been devised.

Filtration can be grouped either as deepbed or surface filtration. Algal biomass is retained on the filter mesh in the state of paste or cake in surface filtration. Upon formation of a thin film of cake on the medium surfaces, the initial film would serve

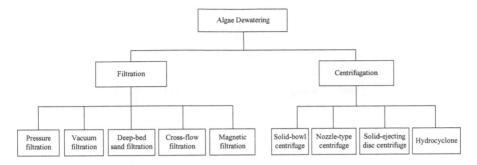

FIGURE 4.3   Process alternatives for algae dewatering.

as a precoat or a secondary filter medium for subsequent deposition of algae. Once the biomass deposits grow thicker, the impedance to pump algal feed across the mesh increases and the filtration output diminishes for a designated pressure drop.

In deepbed filtration, algal biomass is retained within the filter medium. The main setback of using filtration is biofilm fouling or clogging on the medium by the deposited solids. Several methods have been developed to tackle the fouling problem. Frequent filter backwashing is most often included in filtration operations as regular maintenance service to minimize clogging or fouling.

A number of filtration devices have been employed for algae harvesting with different levels of success and are discussed in the following section.

4.5.1.1   Pressure Filtration

Algae can be dewatered through pressure filtration, either by plate-and-frame filtration or pressure chambers mounted with the filter medium. In plate-and-frame filter presses, dewatering is accomplished by pumping the algal fluid across the medium under high pressures. The press, composing a sequence of rectangular plates and recessed on both sides, is fixed face-to-face in a vertical arrangement on a frame mounted with fixed and movable heads. A filter fabric is wrapped over each plate and the plates are secured together and sealed with sufficient force in order to withstand the high pressure exerted during filtration. Algal fluid is pumped into the void between the plates, and pressure is applied and kept for a few hours, forcing the water through the filter fabric and plate outlets. Upon filtration, the plates are opened up and dewatered algal cakes are dislodged and harvested. A complete operation cycle of plate-and-frame filtration is illustrated in Figure 4.4. Coagulant polymers such as polyelectrolytes may be used in conjunction with pressure filtration to enhance the solids content of the dewatered cake.

A variety of pressure chamber designs have been developed which include cylindrical-element filters, vertical-tank-vertical-leaf filters, horizontal-tank-vertical-leaf filters, horizontal-leaf filters, and rotary-drum pressure filters. A comparative study on the applications of various pressure filters for *Coelastrum* harvesting has been conducted (Mohn, 1980). Five varieties of pressure filters, viz. belt press, chamber filter press, cylindrical sieve, pressure suction filter, and filter basket were tested. Solids contents ranging between 5 and 27% of the harvested algae were recorded. The cylindrical sieve, filter basket, and chamber filter press were suggested for algae

FIGURE 4.4   Operation cycle of plate-and-frame filtration.

filtration in terms of energy consumption, process reliability, and filtration capability. The belt filter press was not suggested due to poor cake density, unless prior addition of coagulants was carried out. The pressure suction filter, with low filtration ratio, high investments, and irregular operating costs was also not recommended.

4.5.1.2   Vacuum Filtration

The driving power for vacuum filtration is derived from the suction force applied on the filtrate side across the filter medium. In theory, the pressure drop for vacuum filtration is 100 kPa, but the drop was limited to 70 or 80 kPa in practical operations (Shelef et al., 1984). If large algal particles are predominant in the feed, vacuum filtration could generate a harvested product with solids content as good as that of pressure filtration but at lower operating cost.

Various vacuum filters, namely belt filter, filter thickener, suction filter, and vacuum drum filters precoated/uncoated with potato starch have been examined for the harvesting of *Coelastrum* (Mohn, 1980). Solids contents ranging between 5 and 37% of the harvested product were recorded. In consideration of energy consumption, dewatering capability and reliability, suction filter, precoated vacuum drum filter, and belt filter were recommended. The precoated filter was also used for the harvesting of minute microalgae such as *Scenedesmus* (Shelef et al., 1984). The uncoated vacuum drum filter was not recommended because of inferiority and unreliability due to filter fouling. Filter thickeners were also not suggested because of poor solids contents (3–7%) of the harvested algae, low filtration speed, high energy consumption, and low reliability.

Harvesting of microalgae with a belt filter precoated with pine crafts fibers and eucalyptus was investigated (Dodd and Anderson, 1977). The investigation found that application of the precoat led to escalated costs and detrimental process complexity. In a separate study on vacuum filtration, finely weaved fabric instead of precoated filter showed a relatively low energy requirement while no chemicals were used (Dodd, 1980). It was claimed effective when harvesting large-size algae such as *Micractinium*, but experienced problems of fouling dealing with tiny species like *Chlorella*. The capital investments were pricier than in dissolved-air floatation, but the operating costs were the cheapest among the harvesting methods tested other than natural settling (Dodd, 1980).

4.5.1.3 Deepbed Sand Filtration

In deepbed sand filtration, algal biomass is harvested through a sand medium. Algal fluid flows through the medium and biomass is captured within the sand bed. Sand filtration is usually operated on batch mode. Once the pressure drop exceeds the allowable limit, the operation must be discontinued for filter backwashing.

Sand filtration operated in intermittent mode was studied in a wastewater treatment plant upgrading (Middlebrooks and Marshall, 1974). The study indicated that only large particle size algae could be harvested and tiny algae trapped within the medium could not be effectively harvested. A successful harvesting of algae thriving in pond effluent (mean solids content of 30 mg/L) through intermediate sand filtration was reported (Reynolds et al. 1974). The filtration process, however, encountered severe operating problems of drastic clogging and deterioration in filtration output.

4.5.1.4 Cross-Flow Ultrafiltration

Cross-flow ultrafiltration has been adopted in the treatment of pond effluents for harvesting concentrated algae as animal feed. Satisfactory solids contents of harvested algae had been harvested producing highly clarified treated effluent. The main drawback of this process was the high operating costs due to high energy demands.

4.5.1.5 Magnetic Filtration

Magnetic filtration was originally applied in wastewater treatment for the removal of heavy metallic substances and suspended particles (Bitton et al., 1974). Subsequently, magnetic separation employing magnetic particles (like Fe_3O_4 magnetite) held in suspension was adopted for algae harvesting (Yadidia et al., 1977). Algal biomass and the magnetic beads were mixed and aggregated and the mixture was flowed through a filter medium enclosed by a magnetic field to retain the magnetic aggregates. Algae removal in the range of 55–94% by an industrial magnetic filter augmented with alum coagulant was accounted for (Bitton et al., 1974). Higher removal efficiency of over 90% was reported with Fe_3Cl (5–13 mg/L) as primary coagulant and magnetite as magnetic beads (500–1200 mg/L) for algal pond harvesting (Yadidia et al., 1977).

4.5.2 Centrifugation

Centrifuges are similar to sedimentation tanks except for the driving force used to separate algae. Algal cells held in suspension are separated from the liquid through a centrifugal force which is much higher than the gravity force. Rotating-wall centrifugation vessels are similar to bowl-shaped settlers with their base enfolded around a center axis of rotation. High speed rotation generates a powerful gravitational force a few thousand times higher than the gravity force. Feed is fed into the bowl and the centrifugal force generated by the rotation would spin out the solids in the suspension towards the rotating bowl wall. Clarified supernatant is flowed through a skimming tube or overflow weir, while solids are retained within the bowl in the case of batch operation, or are constantly or intermittently withdrawn in the case of continuous centrifugation.

Different centrifugation systems were studied for algae harvesting (Mohn and Soeder, 1978; Mohn, 1980; Moraine et al., 1980; Shelef et al., 1980, 1984). Some of the systems were found very efficient as a single-step separation, whereas some other systems were deemed inferior or needed prior thickening of the feed. Batch operation of centrifugation was less attractive because it had to be terminated for algae removal. The advantages of some of the centrifugation methods include reliability and efficiency, but high operating expenditures usually offset the merits of such a system for algae harvesting.

Centrifuge devices such as solid-bowl decanter centrifuge, nozzle-type centrifuge, tubular centrifuge, hydrocyclone and solids-ejecting disc centrifuge employed in algae harvesting are discussed in the following sections.

4.5.2.1 Solid-Bowl Decanter Centrifuge

Solid-bowl decanter centrifuge features by a conical-shaped bowl placed horizontally which consists of a screw conveyor that rotates in parallel with the bowl. Slurry is fed through the center shaft and is spun towards the bowl circumference. Spun solids are pulled by the screw conveyor to one side of the bowl for discharge, while clarified liquid forms a concentric inner stratification which flows over an adjustable weir for discharge. The helical screw conveyor that pushes the centrifuged solids operates at a higher rotational velocity than the bowl.

A solid-bowl screw centrifuge system was used to separate different kinds of algae (Mohn, 1980). With an input suspension of 2% solids, the solids content of the harvested algal cake increased to 22%. While the system reliability appeared to be superb, the energy demand was far too exorbitant. An effort to centrifuge an algal suspension of 5.5% solids harvested from a flotation operation by a co-current solid-bowl decanter centrifugation system was futile (Shelef et al., 1980). Consequently, the algal biomass concentration was improved to 21% solids by lowering the screw conveyor rotation to 5 rpm (Shelef et al., 1984). Concurrent application of solid-bowl centrifuge with polyelectrolyte coagulant was recommended for enhancement of harvesting efficiency (Shelef et al., 1984).

4.5.2.2 Nozzle-Type Centrifuge

Continuous harvesting of the algal cake was possible with the nozzle-type disc centrifuge (Shelef et al., 1984). The bowl was adapted such that slurry space contained a conical segment to provide enough storage for a continuous flow of the discharge. The bowl wall surfaces inclined towards a peripheral zone fitted with evenly-spaced nozzles. The quantity and nozzle size were perfected to evade cake deposition and to attain a reasonable concentration of algal cake.

In the application of nozzle-type disc centrifuge, the effects of nozzle size on flow, concentration efficiency, and output slurry solids content were looked into (Golueke and Oswald, 1965). In comparison with other harvesting means, the performance of the nozzle-type centrifuge appeared to be favorable, but it remained less attractive because of high energy consumptions and capital costs. In other investigations, the nozzle-type disc centrifuge seemed to be more efficient to harvest *Scenedesmus* than *Coelastrum* (Mohn and Soeder, 1978; Mohn 1980). By recirculating the underflow back to the feed stream, the solids concentration of the loose suspension (0.1%) was concentrated by a factor of 15–150%. The reliability of this centrifuge can be maintained so long as clogging at the nozzles is averted.

4.5.2.3 Solid-Ejecting Disc Centrifuge

Solid-ejecting disc centrifuge offers cyclic cake discharge by adjusting its valve-regulated peripheral ports through a timer or an automatic triggering setup. The merit of this system was its capability to harvest algal cake in one step without chemical dosing (Mohn and Soeder, 1978; Mohn, 1980; Shelef et al., 1984). Various types of algae were effectively harvested by this centrifuge, producing an algal cake of 12–25% solids (Mohn, 1980; Moraine et al., 1980). The degree of algae separation increased with the liquid retention time (diminishing feed flow), and the harvested cake concentration was influenced by the time gap between successive cake ejection (Shelef et al., 1984). Solid-ejecting disc centrifugation was reported to be very reliable; the only drawback noted was that particles finer than algae may remain in the overflow thereby reducing the separation efficiency (Moreine et al., 1980). Excessive energy and capital costs made this concentration method unappealing.

4.5.2.4 Hydrocyclone

Hydrocyclone operation was not considered a reliable means of algae concentration because solids contents up to 0.4% with a concentration factor of 4 could be achieved (Molina Grima et al., 2003). Compared to other harvesting processes, the advantages of hydrocyclone devices were low capital costs and low energy demand of $0.3\,kwh/m^3$. Nonetheless, hydrocyclones can only process limited strains of algae and their efficiency was highly relying on feed solids content (Origin Oil, 2010). Furthermore, hydrocyclones have been reported to break up natural agglomerates of sea algae *Phaeocystis* (Veldhuis et al., 2006), and may also disrupt algal flocs, thereby making the following harvesting more difficult.

It seemed that hydrocyclones may play a role in algae concentration, confined to pretreatment prior to subsequent dewatering. The potential of centrifugation for biofuel production lied with the use of the entire algal biomass (Milledge, 2010a). One kg of dry algal solids consisting of 20% oil would produce about 1.9 kwh of biodiesel, but the energy potential of the whole biomass is approximately 6 kwh (Milledge, 2010b). Thus, the utilization of the entire biomass is a key consideration in positive energy equilibrium in biofuel production (Heaven et al., 2011; Sialve et al., 2009; Milledge, 2010a; Stephenson et al., 2010).

4.6 CHALLENGES AND PROSPECTS

Algal biofuel is a hotly debated subject as its sustainability remains to be ascertained. By virtue of the dilute algal feed, energy balance and cost effectiveness were two main challenges for the practical production of biofuels (Dismukes et al., 2008; Reijnders, 2008). Algae must be harvested in an economical manner before subsequent drying and oil extraction. The energy requirement for overall harvesting could be significantly reduced if only algae feed could be thickened by as much as 30–50 times through coagulation-flocculation cum gravity sedimentation preceding dewatering (Jorquera et al., 2010; Show et al., 2014).

The majority of the harvesting methods depict a number of drawbacks not just because of the high operating costs. Low efficiencies and poor product qualities also contribute to the shortcomings. Mechanical separation methods such as sedimentation, centrifugation, and filtration may rupture algal cells causing leakage of cellular substance, thus giving rise to deterioration in algal quality. With regard to flocculation, concentrated metal salts used as coagulants could also impair the final product quality (Kim et al., 2005). Even though centrifugation is a very effective harvesting method for algae, the process is exceptionally energy-demanding. In addition, high investment and operating costs plus excessive maintenance obligations are the main liabilities of centrifugation (Molina Grima, et al., 2003; Bosma et al., 2003; Shen et al., 2009; Show et al., 2015).

A comparison of various algae concentration methods outlining their strengths and weaknesses was examined as shown in Table 4.1 (Fasaei et al., 2018). The economic evaluation shows that operating costs and energy consumption are in the range of 0.1–0.6 €/kg algae and 0.1–0.7 kWh/kg algae, respectively, for closed cultivation using mechanical systems. For dilute cultures from open cultivation systems, the operating costs increased to 0.5–2 €/kg algae and the energy consumption to 0.2–5 kWh/kg algae. Despite the higher investment costs of the mechanical systems,

---

TABLE 4.1

Comparison of Algae Concentration Processes Outlining Advantages and Disadvantages

Process	Advantage	Disadvantage
Chemical flocculation	• Low energy demand • Low equipment cost	• Problematic recovery of flocculants
Sedimentation	• Low energy demand • Easy application	• Slow rates • Large operational area • Low recovery • Limited application, suitable for large-size algae
Pressure filtration	• Low energy demand • High recovery	• Discontinuous • Clogging or fouling
Vacuum filtration	• Continuous	• Relative high harvesting cost • Clogging or fouling
Membrane filtration	• Efficient for small-scale processing • High recovery	• Fouling • High capital cost
Centrifugation	• Continuous • Efficient for large-scale processing • High recovery	• High capital cost

Adapted from Fasaei et al. (2018).

the total costs can be reduced by a high level of automation evading much of the labor cost.

At this stage of development, it remains doubtful if the production of biodiesel is environmentally sustainable and which transformation stages require further refinement and optimization. Life cycle assessment (LCA) was conducted to examine the energy equilibrium and the latent environmental impacts of the total algae processing chain (Lardon et al., 2009). Applying a cradle-to-grave approach right from the initial algae cultivation to the final biodiesel combustion, the LCA justified the prospect of algae as an energy supply. However, the assessment underlined the imperative requisite of lowering the energy and fertilizer consumptions. In a separate LCA comparative study examining algal biodiesel production from canola and ultra-low sulfur diesel, in terms of emissions and costs, the necessity for a high production rate appeared to be a key factor in making algal biodiesel commercially viable (Campbell et al., 2011).

The news of biogas energy production from anaerobic treatment of waste oil cakes from oil extraction as a means to reduce energy requirement and to reuse part of the mineral fertilizers is encouraging. This may shed light on tackling the energy and cost challenges of algal biofuel (Lardon et al., 2009). Wastes or by-products produced from existing facilities such as flue gas containing CO_2 from power plants, and effluent containing nutrients such as nitrogen, phosphorous, and other micronutrients from wastewater treatment plants could be transformed into raw material supply for commercial algae cultivation. Use of CO_2 for algal photosynthetic conversion would lead to carbon sequestration, while ingestion of waste nutrients for algal growth would eradicate fossil fuel-based fertilizers, thus mitigating emissions.

A holistic techno-economic assessment is needed to draw a meaningful comparison between algal biofuel and other fossil fuels. On top of tangible benefits that could be quantified from algal biofuel over fossil fuels, intangible benefits such as CO_2 capture, ingestion of waste nutrients-placing fertilizers, and biogas energy produced from anaerobic digestion should also be taken into account. All these benefits could be calculated through a holistic LCA, which might create a possibility for claims of certified emission reductions (or carbon credits) under the Paris Agreement for emissions mitigation. It was reported that the emissions caused by algae production can probably be offset by the emission mitigations deriving from the enhanced production efficiency and sequestration capacity of algae (Williams and Laurens, 2010). It may not be an unrealistic hope that future technological advancement will bring about a global shift towards the production of energy efficient and commercially viable algal biofuel.

4.7 CONCLUSIONS

Algal biofuels appear to be promising fuels in view of their potential to reduce reliance on fossil fuels and to mitigate greenhouse gas emissions. There have been extensive development and improvement in the algal biofuel yield and production rate. For practical applications, the key issues of energy and cost for large-scale production must be tackled. As energy efficiency and cost effectiveness are two major challenges in algae harvesting, R&D work tackling these challenges is to be intensified.

REFERENCES

Adan, B., Lee, E.W. 1980. "High rate algae growth pond under tropical conditions". Presented at a workshop on waste treatment and nutrient recovery. Singapore, 27–29 February.

Azarian, G.H., Mesdaghinia, A.R., Vaezi, F., Nabizadeh, R., Nematollahi, D. 2007. "Algae removal by electro-coagulation process, application for treatment of the effluent from an industrial wastewater treatment plant". *Iran Journal of Public Health* 36, 57–64.

Bare, W.F.R., Jones, N.B., Middlebrook, E.J. 1975. "Algae removal using dissolved air flotation". *Journal of the Water Pollution Control Federation* 47, 153–169.

Benemann, J.R., Kopman, B.L., Weismsman, D.E., Eisenverg, D.E., Goebel, R.P. 1980. "Development of microalgae harvesting and high rate ponds technologies in California". In *Algae Biomass*, B. Shelef and C.J. Solder (eds), Elsevier, Amsterdam, 457.

Benoufella, F., Laplanche, A., Boisdon, V., Bourbigot, M.M. 1994. "Elimination of microcystis cyanobacteria (blue-green algae) by an ozoflotation process – a pilot-plant study". *Water Science and Technology* 30, 245–257.

Betzer, N., Argaman, Y., Kott, Y. 1980. "Effluent treatment and algae recovery by ozone-induced flotation". *Water Research* 14, 1003–1009.

Bitton, G., Mitchell, R. De Latour, C., Maxwell, E. 1974. "Phosphate removal by magnetic filtration". *Water Research* 8, 107–109.

Bosma, R., van Spronsen, W.A., Tramper, J., Wijffels, R.H. 2003. "Ultrasound – a new separation technique to harvest microalgae". *Journal of Applied Phycology* 15, 143–153.

Campbell, P.K., Beer, T., Batten, D. 2011. "Lifecycle assessment of biodiesel production from microalgae in ponds". *Bioresource Technology* 102, 50–56.

Chen, M.Y., Liu, J.C., Ju, Y.H. 1998. "Flotation removal of algae from water". *Colloids and Surfaces B* 12, 49–55.

Cheng, Y.L., Juang, Y.C., Liao, G.Y., Ho, S.H., Yeh, K.L., Chen, C.Y., Chang, J.S., Liu, J.C., Lee, D.J. 2010. "Dispersed ozone flotation of Chlorella vulgaris". *Bioresource Technology* 101, 9092–9096.

Cheng, Y.L., Juang, Y.C., Liao, G.Y., Tsai, P.W., Ho, S.H., Yeh, K.L., Chen, C.Y., Chang, J.S., Liu, J.C., Chen, W.M., Lee, D.J. 2011. "Harvesting of *Scenedesmus obliquus* FSP-3 using dispersed ozone flotation". *Bioresource Technology* 102, 82–87.

Chheda, P., Grasso, D. 1994. "Surface thermodynamics of ozone-induced particle destabilization". *Langmuir* 10, 1044–1053.

Contreras, S., Pieber, M., del Rio, A., Soto, M.A., Toha, J., Veloz, A. 1981. "A highly efficient electrolytic method for microalgae flocculation from aqueous cultures". *Biotechnology and Bioengineering* 23, 1165–1168.

Deirue, F., Seiter, P.A., Sahut, C., Cournac, L., Roubaud, A., Peltier, G., Froment, A.K. 2012. "An economic, sustainability, and energenic model of biodiesel production from microalgae". *Bioresource Technology* 112, 191–200.

Dismukes, G.C., Carrieri, D., Bennette, N., Ananyev, G.M., Posewitz, M.C. 2008. "Aquatic phototrophs: efficient alternatives to land-based crops for biofuels". *Current Opinion in Biotechnology* 19, 235–240.

Divakaran, R., Pillai, V.N.S. 2002. "Flocculation of algae using chitosan". *Journal of Applied Phycology* 14, 419–422.

Dodd, J.C. 1980. "Harvesting algae grown on plg wastes in Singapore", Paper presented at a Workshop on High Rate Algae Ponds held in Singapore.

Dodd, J.C., Anderson, J.L. 1977. "An integrated high-rate pond-algae harvesting system". *Progress in Water Technology* 9, 713–26.

Farvardin, M.R., Collins, A.G. 1989. "Preozonation as an aid in the coagulation of humic substances – Optimum preozonation dose". *Water Research* 23, 307–316.

Fasaei, F., Bitter, J.H., Slegers, P.M., van Boxtel, A.J.B. 2018. "Techno-economic evaluation of microalgae harvesting and dewatering systems". *Algal Research* 31, 347–362.

Friedman, A.A., Peaks, D.A., Nichols, R.L. 1977. "Algae separation from oxidation pond effluents". *Journal of the Water Pollution Control Federation* 49, 111–119.

Golueke, C.G., Oswald, W.J. 1965. "Harvesting and processing sewage grown planktonic algae". *Journal of the Water Pollution Control Federation* 37, 471–498.

Gregory, J. 1977. "Effect of polymers on colloid stability". *Proc. of the NATO Adv. Study on the Scientific Basis of Flocculation*, Cambridge, England, 1.

Habib, M.A., Parvin, M., Huntington, T.C., Hasan, M.R. 2008. "A review on culture, production and use of spirulina as food for humans and needs for domestic animals and fish". FAO Fisheries and Aquaculture Circular No. 1034, Food and Agriculture Organization of the United Nations, FAO Fisheries and Aquaculture Department Rome, Italy.

Heaven, S. Milledge, J., Zhang, Y. 2011. "Comments on 'Anaerobic digestion of microalgae as a necessary step to make microalgal biodiesel sustainable'". *Biotechnology Advances* 29(1), 164–167.

Henderson, R.K., Baker, A., Parsons, S.A., Jefferson, B. 2008. "Characterisation of algogenic organic matter extracted from cyanobacteria, green algae and diatoms". *Water Research* 42, 3435–3445.

Henderson, R.K., Parson, S.A., Jefferson, B. 2009. "The potential for using bubble modification chemicals in dissolved air flotation algae removal". *Separation Science and Technology* 44, 1923–1940.

Jin, P.K., Wang, X.C., Hu, G. 2006. "A dispersed-ozone flotation (DOF) separator for tertiary wastewater treatment". *Water Science and Technology* 53, 151–157.

Jorquera, O., Kiperstok, A., Sales, E.A., Embirucu, M, Ghirardi M.L., 2010. "Comparative energy life-cycle analyses of microalgal biomass production in open ponds and photobioreactors". *Bioresource Technology* 101, 1406–1413.

Kim, S.G., Choi, A., Ahn, C.Y., Park, C.S., Park, Y.H., Oh, H.M. 2005. "Harvesting of *Spirulina platensis* by cellular flotation and growth stage determination". *Letters in Applied Microbiology* 40, 190–194.

Koopman, B.L., Thomson, R., Yackzan, R., Benemann, J.R., Oswald, W.J. 1978. *Investigation of The Pond Isolation Process for Microalgae Separation from Woodlands Waste Pond Effluents*. Final Report, U.C. Berkeley.

Kumar, H.D., Yadava, P.K., Gaur, J.P. 1981. "Electrical flocculation of the unicellular green algae *Chlorella vulgaris*". *Aquacul. Bot.* 11, 187–195.

Lardon, L., Helias, A., Sealve, B. Steyer, J.P., Bernard, O. 2009. "Life-cycle assessment of biodiesel production from microalgae". *Environmental Science and Technology* 43, 6475–6481.

Lee, D.J., Liao, G.Y., Chang, Y.R., Chang, J.S. 2012. "Coagulation-membrane filtration of Chlorella vulgaris". *Bioresource Technology* 112, 184–189.

Levin, G.V., Clendenning, J.R., Gibor, A., Bogar, F.D. 1962. "Harvesting of algae by froth flotation". *Applied Microbiology* 10, 169–175.

Liu, W.X., Clarens, A.F., Colosi, L.M. 2012. "Algae biodiesel has potential despite inconclusive results to date". *Bioresource Technology* 104, 803–806.

McGarry, M.G., Durrani, S.M.A, 1970. *Flotation as a Method of Harvesting Algae from Ponds*. Research program report No.5. Asian Institute of Technology, Bangkok.

Middlebrooks, E.J., Marshall, G.R. 1974. In *Upgrading Wastewater Stabilization Ponds to Meet New Discharge Requirements* Utah, Water Res. Lab., Utah State Univ., Logan, PRwF 159–1.

Milledge, J.J. 2010a. "The challenge of algal fuel: Economic processing of the entire algal biomass". *Condens. Matter Mater. Eng. Newsl.* 1(6), 4–6.

Milledge, J.J. 2010b. "The potential yield of microalgal oil". *Biofuels International Magazine* 4(2), 44–45.

Mohn, F.H., Soeder, C.J. 1978. "Improved technologies for harvesting and processing of microalgae and their impact on production costs". *Arch. Hydrobiol. Bech. Ergebn. Lemnol.* 11, 228–253.

Mohn, F.H. 1980. "Experiences and strategies in the recovery of biomass from mass cultures of microalgae". In *Algae Biomass*, B. Shelef and C.J. Solder (eds), Elsevier, Amsterdam, 547–571.

Molina Grima, E., Belarbi, E.H., Acién Fernández, F.G., Robles Medina, A., Chisti, Y. 2003. "Recovery of microalgal biomass and metabolites: Process options and economics". *Biotechnology Advances* 20, 491–515.

Moraine, R., Shelef, G., Sandbank, E., Bar Moshe, Z. and Schwarbard, L. 1980. "Recovery of sewage borne algae: Flocculation and centrifugation techniques". In *Algae Biomass*, G. Shelef and C.J. Solder (eds), Elsevier, North Holland.

"Origin Oil 2010. Algae harvesting, dewatering and extraction". Paper presented at the World Biofuel Markets, Amsterdam.

Pearsall, R.V., Connelly, R.L., Fountain, M.E., Hearn, C.S., West, M.D., Hebner, R.E., Kelley, E.F. 2011. "Electrically dewatering microalgae". *IEEE Transactions on Dielectrics and Electrical Insulation* 18, 1578–1583.

Phoochinda, W., White, D.A. 2003. "Removal of algae using froth flotation". *Environmental Technology* 24, 87–96.

Phoochinda, W., White, D.A., Briscoe, B.J. 2005. "Comparison between the removal of live and dead algae using froth flotation". *J. Water SRT-AQUA* 54, 115–125.

Plummer, J.D., Edzwald, J.K. 2002. "Effects of chlorine and ozone on algal cell properties and removal of algae by coagulation". *J. Water SRT-Aqua* 51, 307–318.

Reckhow, D., Singer, P., Trussel, R.R. 1986. "Ozone as a coagulant aid". In *Ozonation: Recent Advances and Research Needs. AWWA Seminar Proc. No. 20005*, Denver, CO, Am. Water Works Assoc., 17–46.

Reijnders, L. 2008. "Do biofuels from microalgae beat biofuels from terrestrial plant?" *Trends in Biotechnology* 26, 349–350.

Reynolds, J.H., Harris, S.E., Hill, D., Felip, D.S., Middlebrooks, E.J. 1974. *Intermittent Sand Filtration to Upgrade Lagoons Effluent.* Preliminary report in Upgrading Wastewater Stabilization Ponds to meet new Discharge Standards by Middlebrooks E.J. et al, Utah Water Res. Lab. Logan.

Reynolds, J.H., Middlebrooks, E.J., Porcella, D.B., Grenney, W.J. 1975. "Effects of temperature on oil refinery waste toxicity". *Journal of the Water Pollution Control Federation* 46, 2674–2693.

Sandbank, E., Shelef, G., Wachs, A.M. 1974. "Improved electroflotation for the removal of suspended solids from algae pond effluents". *Water Research* 8, 587–592.

Sandbank, E. 1979. "Harvesting of microalgae from wastewater stabilization pond effluents and their utilization as a fish feed". D.Sc. thesis presented to the senate of the Technion – Israel Institute of Technology.

Schlesinger, A., Eisenstadt, D., Bar-Gil, A., Carmely, H., Einbinder, S., Gressel, J. 2012. "Inexpensive non-toxic flocculation of microalgae contradicts theories; overcoming a major hurdle to bulk algal production". *Biotechnology Advances* 30(5), 1023–1030.

Shelef, G., Azov, Y., Moreine, R., Oron, G. 1980. "Algae mass production as an integral part of a wastewater treatment and reclamation system". In *Algae Biomass*, B. Shelef and C.J. Solder (eds), Elsevier, North Holland.

Shelef, G., Sukenik, A., Green, M. 1984. "Microalgae harvesting and processing: A literature review". Report prepared for the US Department of Energy, Technion Research and Development Foundation Ltd. Haifa, Israel.

Shen, Y., Yuan, W., Pei, Z.J., Wu, Q., Mao, E. 2009. "Microalgae mass production methods". *Transactions of the ASABE* 52, 1275–1287.

Sialve, B., Bernet, N., Bernard, O. 2009. "Anaerobic digestion of microalgae as a necessary step to make microalgal biodiesel sustainable". *Biotechnology Advances* 27(4),
409–416.

Show, K.Y., Lee, D.J., Chang J.S. 2013. "Algal biomass dehydration". *Bioresource Technology*
135, 720–729.

Show, K.Y. and Lee, D.J. 2013. "Algal biomass harvesting". In *Biofuels from Algae*, A. Pandey,
D.-J. Lee, Y. Chisti and C.R. Soccol (Eds.), Elsevier, North Holland, pp. 85–110.

Show, K.Y., Lee, D.J., Tay, J.H., Lee, T.M., Chang, J.S. 2014. "Microalgal drying and cell
disruption – Recent advances". *Bioresource Technology* 184, 258–266.

Show, K.Y., Lee, D.J., Mujumdar, A.S. 2015. "Advances and challenges on algae harvesting
and drying". *Drying Technology* 33(4), 386–394.

Stephenson A.L., Kazamia E., Dennis J.S., Howe C.J., Scott S.A., Smith A.G. 2010. "Life-
cycle assessment of potential algal biodiesel production in the United Kingdom: A comparison of raceways and air-lift tubular bioreactors". *Energy Fuels* 24(7), 4062–4077.

Sukenik, A., Shelf, G. 1984. "Algal autoflocculation—Verification and proposed mechanism". *Biotechnology and Bioengineering* 26, 142–147.

Svarovsky, L. 1979. "Advanced in solid-liquid separation II sedimentation, centrifugation and
flotation". *Chemical Engineering Journal* 16, 43–105.

Tenney M.W., Echelberger W.F., Schuessler R.G., Pavpni J.L. 1969. "Algal flocculation with
synthetic organic polyelectrolytes". *Applied Bacteriology* 18, 965–971.

Veldhuis M.J.W., Fuhr F., Boon J.P., Ten Hallers-Tjabbers C. 2006. "Treatment of ballast
water; how to test a system with a modular concept?" *Environmental Technology* 27(8),
909–921.

Wang, B., Li, Y., Wu, N., Lan, C.Q. 2008. "CO_2 bio-mitigation using microalgae". *Applied
Microbiology and Biotechnology* 79, 707–718.

Wettman, J.W., Cravens, J.B. 1980. "Cost effective lagoon upgrading with microscreens".
Proc. the 3rd Ann. Poll. Cont. Assoc., Oklahoma, June 5.

Williams, P.J., Laurens, L.M. 2010. "Microalgae as biodiesel and biomass feedstocks: Review
and analysis of the biochemistry, energetics and economics". *Energy & Environmental
Science* 3, 554–590.

Yadidia, R., Abeliovich, A., Belfort, G. 1977. "Algae removal by high gradient magnetic
filtration". *Environmental Science and Technology* 11, 913–916.

Zhang, G.M., Zhang, P.Y., Fan, M.H. 2009. "Ultrasound-enhanced coagulation for Microcystis
aeruginosa removal". *Ultrasonics Sonochemistry* 16, 334–338.

Zheng, H.L., Gao, Z., Yin, J.L., Tang, X.H., Ji, X.J., Huang, H. 2012. "Harvesting of microalgae by flocculation with poly (gamma-glutamic acid)". *Bioresource Technology* 112,
212–220.

Zhou, W., Cheng, Y., Li, Y., Wan, Y., Liu, Y., Lin, X., Ruan, R. 2012. "Novel fungal pellet-
ization-assisted technology for algae harvesting and wastewater treatment". *Applied
Biochemistry and Biotechnology* 167(2), 214–228.

5 Drying of Algae

Y.C. Ho, K.Y. Show, Y.G. Yan, and D.J. Lee

CONTENTS

5.1 INTRODUCTION

Algae are a highly diverse group of organisms with important functions in aquatic habitats. They are also known as photoautotrophic organisms with unicellular reproductive structures, chlorophyll *a*, and a thallus, which do not have roots, stems, or leaves. There are 4–13 algal divisions, 24 classes, and about 26,000 species according to several taxonomic schemes (Raven and Johnson, 1992). Algae could be divided by size into two groups, (1) microalgae which are microscopic organisms with size ranging from 1 to 50 µm, i.e. Chlorella vulgaris, Microcystis aeruginosa, Asterionella formosa, Melosirasp, Micractinium, Scenedesmus, Spirulina, Oscillatoria, and Synechocystis, and (2) macroalgae known as "seaweed," i.e. Undaria, Porphyra, Euchema, and Gracilaria. Moreover, researchers explored the possibilities of polar snow algae (Chlamydomonas pulsatilla) (Hulatt et al., 2017). Milledge et al. (2014) reviewed the production of fuels from microalgae at large. Further, limited studies have been conducted on the production from macroalgae.

Algae were first investigated as a source for biofuel production in the 1950s. The investigation of a new energy source was strongly demanded during the oil crisis in the 1970s. Lately, the utilization of microalgae in biofuel production has caught much attention due to the high constituent of lipids content which can be used as feedstock to achieve high biofuel production rate. Compared to other biofuel production derivatives, microalgae have a high growth rate (100 times faster than terrestrial plants), high biomass production, and low land use (Sanyano et al., 2013). In addition, Lam and Lee (2014) noted that an average of 4.5–7.5 tonne/ha/year of crude microalgae lipids production rate could be achieved with an assumption that 30% of microalgae biomass is lipids content. In addition to the contribution in biofuel production, residue of microalgae biomass, which are poor in lignin but rich in proteins and other compounds of commercial interest, can be further processed and utilised in animal feed production and biorefinery-based production, for example, nutritional supplements, cosmetics, and pharmaceutical products (Barros et al., 2015). Another advantage of microalgae growth is the assimilation of carbon dioxide in the atmosphere and flue gases by algae during photosynthesis (Lam and Lee, 2012; Show et al., 2013).

Currently, biodiesel production derived from algae goes through different processes, i.e. algal strain development and cultivation, harvesting, and subsequent processes such as dewatering, drying, oil extraction, and fractionation (Zheng et al., 2012).

One of the early studies performed by Shelef et al. (1984) discussed the methods of harvesting microalgae. In the study, it noted the microalgae harvesting technologies during the years did not show any revolutionary conceptual advances. However, the authors revealed that by establishing dewatering processes, solids requirements up to 30% could be obtained. Further, algae drying methods are also required for more concentrated solids. There are various drying methods for algae and different techniques influenced by the scale of microalgae production. In principle, drying methods and cell disruption techniques are two predominant criteria for effective biofuel production from microalgae (Guldhe et al., 2014). Drying operation systems are different in view of capital investment and energy requirements. All in all, the drying method selection depends on the operation scale and the intended dried product usage. For instance, the sun drying method is economical for large production scale. However, it requires large land space.

Algae drying is a complicated transient operation of simultaneous heat and mass transfer with various rate processes. Liquid from wet material is evaporated with the heat supply during the drying process. The drying rate is determined by the content of moisture and the material temperature to be dried, the humidity, and the velocity of air in contact with the material. Every material behaves differently during drying, depending on their moisture content. In the end, algae drying is an energy-intensive process and a huge amount of energy is consumed for biomass drying for economic use (Kaliyan and Vance Morey, 2009). Adams et al. (2011) noted the drying process involves the removal not only of free water but also of the water between particles and of chemically bound water in the structure. Researchers have tried different methods, which include natural, technology-aided, and equipment-aided approaches for algae drying. Table 5.1 shows the categories and state-of-the-art methods for algae drying.

This review aims to provide insights on effective methods for the drying of algae for future developments in the use of microalgae in biofuel production. Challenges and prospects of algae drying are outlined as well.

TABLE 5.1

Categories and Examples of Algae Drying Methods

Methods	Solutions
Stabilization	Ozonation, polyelectrolytes charge, oxidation
Natural	Sun drying
Equipment-aided	Freeze dry, spray dry, rotary, microwave, oven, drum
Technology-aided	Heat circulation technology, membrane, osmotic dehydration

5.2 DESTABILIZATION OF ALGAE

The selection of a drying method impacts the stability of microalgae (Vandamme, 2013). Thus, it is essential to understand the mechanism in the destabilisation process which may lead to settling and, subsequently, reduction in water content.

Algae surface charge, size, and cell density are associated with the stability of microalgae in growth medium, which influences their separability from aqueous suspensions (Tenney et al., 1969; Show et al., 2013). The electric interactions between algae cells and cell interactions with the surrounding culture broth could contribute to the stability of the algae suspension. Further, size and algae cells determine their settling rate, which is important to consider for separation through sedimentation. Precisely, microalgae stabilise in water with small size and a density similar to that of water. Also, microalgae are hydrophilic and have a negatively charged surface, which prevents them from self-aggregation, thus keeping them in suspension (Gerde et al., 2014; Laamanen et al., 2016; Sanyano et al., 2013).

Ozonation affects particles surface charge and their stability (Chheda et al., 1992) and mechanisms for enhanced coagulation of suspended particles (Plummer and Edzwald, 2002). Show et al. (2013) noted that appropriate treatment of ozonation could destabilise the particles, leading to improvement in separation from the medium.

Destabilisation of colloidal suspension like in an algae culture is attainable by the presence of organic polymer by specific interactions at the ionic or atomic levels (Shelef et al., 1984). Commercial polymers with high molecular weights such as polyelectrolytes or polyhydroxy complexes are proven to be superior coagulants or flocculants. Polymeric coagulation-flocculation is explained by the bridging model, which suggests that a polymer could attach itself to the algae particle surface by several segments with the remaining segments extended into solutions. The segments are able to bridge on vacant sites of other algae particles, forming three-dimensional floc networks that could enhance the settleabilities of the particles (Gregory, 1977).

Pre-oxidation is critical in the improvement of algae destabilisation. Thus, Lin et al. (2015) investigated the oxidation effect with sodium hypochlorite (NaOCl) and chlorine dioxide (ClO_2) on the cell integrity of algae. The result showed that pre-oxidation with ClO_2 works better in the destabilisation of particles.

Principles in destabilisation and current densities were investigated to understand the feasibility of employing the electroflotation-oxidation process using a pair of boron-doped diamond (BDD) and aluminum electrodes with green microalgae (Scenedesmus quadricauda) (Ryu et al., 2018). The efficiency was analysed at various current densities. The researchers showed that the greater positive charge could exhibit better destabilisation of the negatively charged surface of algal suspensions through neutralization of charge (Vandamme et al., 2011).

Besides, fungi have the special capability to produce a wide range of extracellular enzyme complexes, including cellulase and xylanase (Kaushik et al., 2013). These could be used as pretreatment of algae before cell disruption. Researches on pretreatment have been conducted using commercial enzymes such as a-amylase, mylogluccosidase (Choi et al., 2010), cellulase, xylanase, lipase, and protease. Prajapati et al. (2015) investigated two fungal strains, namely, Aspergillus lentulus and Rhizopus oryzae, which were isolated from textile wastewater as they were reported to produce good cellulase and xylanase activity. Both are postulated to be essential in algae biomass treatment and cell wall disruption as well. Due to their high cellulose content, algae cell walls frequently contain xylose, which forms the skeletal polysaccharides arranged as xylans microfibrils. In the study, the fungi were used as pretreatment for unicellular microalgae Chroococcus sp.1. The result showed that the release of cellular content through enzymatic action on microalgae cell walls could facilitate the aggregation of microalgae cells. Better pretreatment efficacy (at a loading equivalent to ≈ 21.125 FPU/g algae biomass) was achieved using A. lentulus crude enzyme. Apart from the promising result from the investigation, this could be a solution to improve the technical and economic aspect of biofuel production processes with algae.

The settling velocity of algae cells could be increased by cell dimensions increment. The principle is applied in algae separation processes where chemical coagulants are added to form large algae flocs which will settle at a rapid rate to the reactor or tank bottom. Air bubbles that may attach to the already formed algae flocs will reduce the density of floc, causing it to float atop the vessel. The increase in gravitational force increases the settling velocity of algae cells which is possible by applying centrifugal forces on suspensions of algae (Show et al., 2013).

5.3 DEVELOPMENT IN MICROALGAE DRYING TECHNOLOGY

After algae separation and harvesting, drying of algae will be conducted for biofuel production. Algae drying is a process that dehydrates algae into dewatered algae cakes (sludge), which do not behave as a liquid and can be handled manually or mechanically. The dewatering or drying technique was estimated to utilise 50% and above of the total energy consumption in microalgae processing (Venteris et al., 2014), or around 25% of microalgae energy content (Xu et al., 2011). Nevertheless, the process is challenging and requires an innovative solution. The most feasible methods for the drying of algae should eliminate algae quality degradation from the process. Often, algae are considered dried when their moisture content is less than 10% (Villagracia et al., 2016). Therefore, various methods for the drying of algae for algae biomass or biodiesel production are being studied in order to discover the efficiency of each operational method, as shown in Figure 5.1, and will be discussed in the following sections.

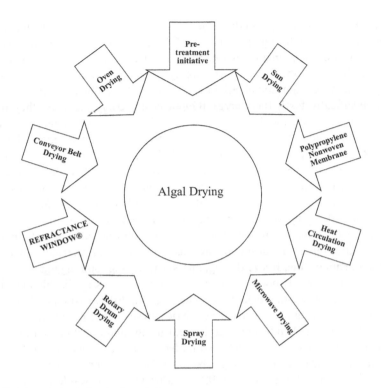

FIGURE 5.1   Methods for drying of algae.

5.3.1 Pretreatment Initiative

Most microalgae have a resistant cell wall, which makes them recalcitrant due to the presence of complex biopolymers (microfibrillar polysaccharides, matrix polysaccharides, and proteoglycans). Pretreatment of algae biomass techniques has been proven to be successful in improving the disintegration of organic substances (Carrère et al., 2010), which made them necessary for microalgae cell disruption and biofuel production prior to algae drying (Chen and Oswald, 1998).

Chen et al. (2014) investigated the effects of different pretreatments on ash reduction and thermal decomposition of wastewater algae by comparing single-stage (centrifugation) and two-stage (centrifugation and ultrasonication) pretreatments. In what follows, the pretreated algae biomass was used to determine the activation energy of thermal decomposition. It was noted that two-stage pretreatments improved the thermal decomposition behaviour of wastewater algae and, subsequently, enhanced the efficiency of bio-crude oil conversion by increasing bio-crude oil yields.

5.3.2 Sun Drying

Algae must be processed as soon as they are harvested. The main reason is to prolong the usability of algae for the next application. Sun drying is one of the options (Prakash et al., 1997). The other available options are low-pressure shelf drying, spray drying, drum drying, fluidised bed drying, freeze drying, and Refractance Window® technology drying (Brennan and Owende, 2010). Among them, sun drying is the most conventional drying method. Innovation drives researchers to seek alternative drying methods which are highly efficient, low cost, environmentally friendly, and have high social acceptance. Seemingly, other methods consume energy, specifically fossil fuel energy (Gupta et al., 2011). As such, they become inefficient and costlier for the biomass-to-liquids (BTL) value chain. Sander and Murthy (2010) noted that thermal algae drying consumed about 3556 kJ/kg fossil fuel-based energy. Further, Lardon et al. (2009) reported that the energy used for algae drying was up to 85% of the total energy consumed in algae processing. When fossil fuel prices escalate, the consumption of energy becomes a major concern. Thus, efforts are poured in to look for alternatives (Aziz et al., 2013). The focus is on using natural drying methods, namely, sun or solar drying methods. They have been applied for a long time with natural convection due to solar radiation. Jully Tan et al. (2016) reported that sun drying was the best drying alternative when compared with drum drying, freeze drying, and spray drying. Sun drying is potentially economical; however, it should be noted that sun drying is applicable and suitable in places with good weather conditions (Zhang et al., 2014). Brink and Marx (2013) investigated the drying of cyanobacterium Microcystis aeruginosa which was found in the Hartbeespoort Dam throughout the year using sand filtration and followed by sun drying. The research was conducted under the average daytime temperature of 24.0°C. It was found that the microalgae biomass showed caloric value at 21.02 MJ/kg which makes it possible to produce 9938 GJ/ha yr of renewable energy with an assumption of 300 effective drying days per year. In principle, solar radiation supplies energy to evaporate the moisture and can reduce the energy requirements of the BTL process. However, Show et al. (2013) noted that solar radiation is uncontrollable and may cause algae overheating. That way, it can destroy the intrinsic properties of the algae, thus the reliability of adopting this method is low and weather dependent.

Guldhe et al. (2014) evaluated and compared the performance of three drying methods, i.e. freeze drying, oven drying, and sun drying of Scenedesmus sp. The authors noted the effects of different drying methods and evinced no significant difference between them. Table 5.2 shows the influence of different drying methods on lipids yield with two types of lipids extraction methods, namely, microwave-assisted and sonication-assisted extraction. It is noteworthy that freeze drying is favourable as it retains the original intrinsic properties of microalgae (C. vulgaris was used in this study) lipids (Widjaja et al., 2009). Further, the heating on localised conditions for microwave- and sonication-assisted conditions is illustrated in Figure 5.2.

In the evaluation of suitable methods of drying, the authors considered time factors and operational costs which include energy consumption. In view of these, sun drying has minimum cost but requires a longer time, i.e. 72 h, as compared to oven drying (12 h) and freeze drying (24 h). Intensive energy was required for oven and freeze drying, with 6 kW and 21.96 kW respectively, as well as sophisticated equipment. Thus, it is suggested that the efficiency of drying methods is as follows: sun drying < oven drying < freeze drying (Guldhe et al., 2014).

However, considering high energy consumption is required for drying methods such as freeze drying, oven drying, and spray drying; they may not be feasible

TABLE 5.2

Lipids Yield by Freeze, Oven, and Sun Drying with Microwave-Assisted and Sonication-Assisted Solvent Extraction

Extraction of Lipids	Drying Method	Dry Cell Weight, Lipid/g
Microwave-assisted solvent extraction	Freeze drying	$29.65 \pm 1.05\%$
	Oven drying	$28.63 \pm 0.42\%$
	Sun drying	$28.33 \pm 1.37\%$
Sonication-assisted solvent extraction	Freeze drying	$19.85 \pm 0.35\%$
	Oven drying	$18.8 \pm 0.1\%$
	Sun drying	$18.9 \pm 0.5\%$

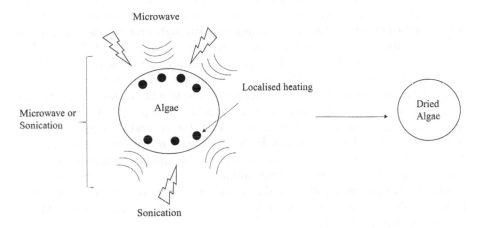

FIGURE 5.2   Drying of algae with microwave- and sonication-assisted condition.

for biofuel or animal feed production (Syed et al., 2016). If land use is taken into account, sun drying may not be feasible as it requires large land areas. Thus, careful selection in the adoption of drying methods is necessary in countries which have land constraints.

Poblete et al. (2018) investigated the solar drying method. The authors fabricated a solar dryer which was made up by a solid concrete base, a polycarbonate front and lateral walls, and a wooden back wall. The roof of the solar dryer was made from glass and had 85% optical transmittance and 0.8 W/m °C thermal conductivity transmittance. The authors noted that a closed solar dryer system was more efficient than an open-air system. This is evinced from the higher degree of moisture evaporation in the closed system, i.e. 86.1%, compared to 67.6% in the open-air system. Also, it is essential to use an air fan to achieve high drying efficiency. Nonetheless, more solar dryer systems could be installed in order to increase the loading capacity.

Apart from sunroof solar systems, other solar drying systems used in the agricultural sector could also be applied to algae drying, i.e. hybrid solar dryer, greenhouse solar dryer, tent solar dryer, cabinet solar dryer. Hybrid solar dryers integrate solar energy with other sources of energy such as biomass or fossil fuel. This dryer is capable to reduce drying time. However, it may rely on the fuel supply. Thus, this method is expensive to operate, especially in countries where fuel energy is in scarce (El Hage et al., 2018).

The greenhouse dryer is commonly used for drying purposes. It is simple to design and operate, and cheap. The walls and roof of the dryer were made up of glass or polycarbonate sheets to enhance solar irradiation penetration. There are two types of greenhouse dryers, i.e. (1) dome, and (2) roof even types. The dome type maximised the solar radiation while the roof even type enhanced air mixing in the dryer (Prakash and Kumar, 2014). It can also be integrated with the PV/T module. Singh et al. (2018) reviewed the solar greenhouse system. The authors concluded this integrated system is suitable for areas with limited electricity.

An interesting study was conducted by Slager et al. (2014) to investigate the combined production of tomato and algae in a greenhouse. In this study, a Matlab-based model was developed to determine productivity and economic feasibility. However, it was found that the dry matter (DM) per unit cost was higher for separated units compared to combined production. The production cost is calculated as €11/kg algae DM and €70/kg. It was higher than in the previous assumption on algae price of €50/kg.

On another note, there are different types of setup for solar dryers, e.g. tent solar dryer and cabinet solar dryer. The former used wooden pillars covered with plastics sheet. The background of plastic was black in colour to better promote the adsorption of solar radiation. The air flowed into it from the top and the algae were dried at the bottom. Tent dryer was relatively cheap and easy to construct. It is noted that the time it took to dry with a tent solar dryer was the same as open-air drying. The only drawback was the fragility towards strong winds. Next, cabinet solar dryer. This is a large box constructed with metal or wood. Shelves were installed in the space between walls. It is worth to mention that a heat storage system can be installed in it as well (El Hage et al., 2018).

Eltawil et al. (2018) investigated the hybrid solar tunnel dryer (STD) with PV system and flat plate solar collector for drying purposes. The aim of the STD design was to collect solar energy through a solar collector, while the PV system aimed to circulate the hot mass of air from flat plate solar collector to SDT via an insulated tube. Thus, the indirect heat dried the sample.

Heat pumps technology could be taken into consideration. Renewable sources were utilised for heat generation and it replaced fuel as energy driver. Also, it is recommended due to economic reasons and thermodynamic performance. However, the drawback of using a heat pump resulted from its operational effects (Ommen et al., 2014). Combined systems can be adopted in order to increase efficiency and reduce the effects derived from the operations. For example, Dott et al. (2012) integrated solar energy and heat pumps systems using direct and indirect solar irradiation.

5.3.3 OVEN DRYING

Oven drying of algae is one of the most common drying methods as it is easily available. Numerous researches for biofuel production adopt the oven drying method (Balasubramanian et al., 2011). However, if it is applied in large scale, for example, industrial scale, it may not be economical and practically feasible. Yet, the research on using the oven is still going on. Abomohra et al. (2014) studied the production of biofuel from Scenedesmus obliquus. The drying of microalgae was performed by using an oven at different temperature settings. The temperature was found to influence the concentration of extracted esterified fatty acids (EFA), but not individual fatty acid proportions. The extracted EFA concentration increased by 6% and 12%, at 75°C and 100°C respectively. At 100°C, the time used for drying was reduced and the fatty acid composition showed no changes. Conversely, Syed et al. (2016) noted drying operating conditions influenced the chemical properties of biomass. Not only that, but physical properties such as size, thickness, breakability, true density, and bulk density depend on drying methods. From an economic perspective, oven drying is more cost effective in terms of energy consumption.

5.3.4 HEAT CIRCULATION DRYING

Drying of algae requires heat for the removal of water for subsequent processes, e.g. cell disruption and transesterification. Careful consideration and delicate technology are necessary to achieve a good quality of dried algae and, subsequently, the end product. González-Fernández et al. (2012) conducted a study on thermal pretreatment and highlighted that at 90°C, it caused greater damage to the cell wall of microalgae biomass (Scenedesmus sp.). Apart from the considerations on temperature, the technology involved focuses on algae drying. Aziz et al. (2013) proposed a novel drying process for brown algae (Laminaria japonica), which is based on heat circulation technology. It employs the principles of exergy elevation and heat pairing for both sensible and latent heat. Interestingly, with this heat circulation technology, the drying process showed a significant reduction in the energy required, up to 90% compared to that used in conventional heat drying, which was based on heat cascade-utilisation technology. Also, the temperature–enthalpy results showed the hot and cold streams curves were almost parallel, resulting in the minimization of exergy losses.

Subsequently, Aziz et al. (2015) studied the integrated system which consists of drying, gasification, and combined cycle by utilising principles of exergy recovery and process integration. In this system for the drying of algae, equipment includes a preheater, main dryer, and superheater. Apparently, a steam tube rotary dryer was adopted for effective evaporation process. Further, a steam tube with an internal heat exchanger is deemed to be a promising drying method. The rotary dryer was agitated at a constant rate and with a slope to foster the discharge by gravity force. Gasification was achieved by adopting a fluidised bed type gasifier as it brought benefits, i.e. carrying of high mass, better heat transfer, efficiency, and more. It is noted that the first exergy elevation occurred when purged steam was compressed at the compressor leading to a high exergy rate of the hot stream. Here, a combination of electrical and thermal energy occurs; therefore, it could elevate the exergy rate of thermal energy. The second exergy elevation occurred in gasification. The steam was superheated by hot gas. It is a combination of energies which results in an elevation of the exergy rate of thermal energy. Nonetheless, around 40% of energy could be harvested, and it led to high energy efficiency, which served as a good alternative.

5.3.5 MICROWAVE DRYING

The microwave is a novel attempt for drying algae without damaging them. It is postulated to have a short drying time. Chlorella vulgaris was used to investigate different microwave intensity levels (300, 600, 900 W) and weights (10, 20, 30 g) at a frequency of 2,450 MHz. The results showed that at a setting of 20 W/g microwave drying condition, it required lower specific energy and, at the same time, maintained high lipids content (Villagracia et al., 2016). Moreover, the energy used in microwave drying was lower than in other drying methods (Bennion et al., 2015; Minowa and Sawayama, 1999; Xu et al., 2011; Villagracia et al., 2016) as shown in Figure 5.3. Obviously, spray drying required the highest energy use.

5.3.6 SPRAY DRYING

Spray drying is known to be used only in highvalue products, such as nutritious products, drugs, or for medicinal use. Also, it is suitable for materials which have high protein content. However, it is relatively expensive and may lead to deterioration of some algae pigments (Desmorieux and Decaen, 2005). Besides, it is deemed feasible to apply it for Dunaliella salina in order to retain rich β-carotene (no change in isomer composition). The authors noted that salt concentration in algae should be reduced for algae drying to be more economical. It could be reduced by centrifugation to achieve 1–2% salt without disrupting the cells; however, this solution may not be suitable for industrial scale (Leach et al., 1998). In

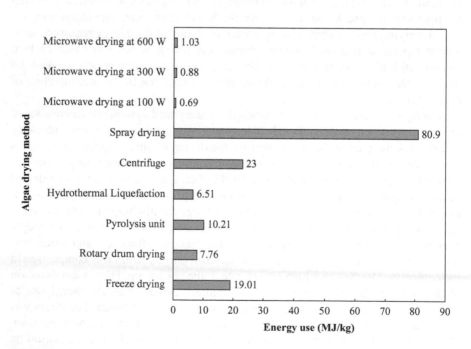

FIGURE 5.3 Energy use in different algae drying methods.

spray drying of D. salina, it is reported that temperature influenced the nature of the polymer matrix as shown in the aspect of permeability of oxygen during storage. In practice, during this drying process, water continuously evaporated through the drying matrix. When the temperature continued to rise, pressure built up and then released water vapour which led to a more open-voided structure. In this way, algae are exposed to oxygen and, subsequently, degradation is gradually greater. Thus, it is suggested that algae incapsulated powder form should be used, as this exhibited milder degradation and the proportion of isomers remained unchanged.

5.3.7 ROTARY DRUM DRYING

Back in 1978, Mohn and Soeder (1978) examined and compared the advantages between spray drying and drum drying for microalgae. Notably, drum drying exhibited better digestibility, fewer energy requirements, and lower investments (Mohn and Soeder, 1978; Show et al., 2015).

Wahlen et al. (2017) reported on the drying of algae (Scenedesmus sp.) by rotary drum drying. The research was conducted with a blend of other terrestrial biomass such as ground pine, sorghum, corn stover, sieved sand, and dried algae. The bench-scale rotary drum dryer was set to an internal temperature of 50°C. When comparing raw algae biomass drying to its blend, the results indicated two factors influencing the drying rate, i.e. moisture content and lignocellulosic, in the biomass. However, the biochemical composition of the biomass showed little effect and thus it is necessary to determine the optimum ratio of algae to terrestrial biomass. Further, the authors concluded that with the presence of other biomass, and at the optimal ratio, the drying rate marked higher than that of algae alone. Biomass blended with algae may applicable when high throughput was required in a rotary drum unable to dry algae directly, or to preserve excess algae during low demand season. Another advantage of blended algae includes lower drying cost. The blends with algae contained 20%, 40%, 60%, and 80% of algae. The result showed that, in 1.7 h, 20% algae blend could be dried completely (0% moisture) and no significant results for 80% algae blend and 100% algae were apparent. Also, at 40% algae, the moisture content could be lowered to 2% with only half of the time required when algae were dried alone using the rotary drum drying method. Figure 5.4 shows a rotary drum and how it works.

5.4 REFRACTANCE WINDOW®

The Refractance Window® system is a drying method used primarily for food technology or food industry. It turns biomaterials into powders, flakes, or sheets. It is commonly used for commercial production of high carotenoid-containing algae. It dries materials in a short period of time, ranging from 3 to 5 min, and retains their properties. This drying system is deemed as cheap compared to freeze drying. In this system, thermal energy is transferred from hot water to where the algae are spread thinly on a plastic conveyor belt. Further, this patented drying system operates at atmospheric pressure (Nindo and Tang, 2007).

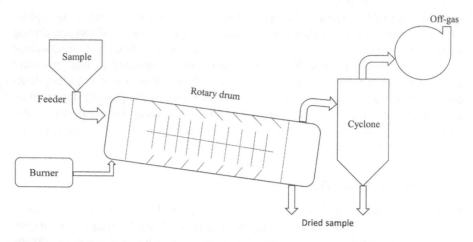

FIGURE 5.4 Rotary drum dryer.

5.4.1 CONVEYOR BELT DRYING

Often, the right dryer for a material is essential to meet good evaporation rate and, at the same time, minimising the thermal damage to microalgae (Esquivel et al., 1993). To preserve properties or composition, freeze drying could be an excellent option; however, it is expensive due to the vacuum condition during drying, which consumes energy (Liapis and Bruttini, 2014). Hosseinizand et al. (2017) analysed the cost of using conveyor belt drying at an industry site. The microalgae were spread on a horizontal belt and the airflow above the belt was not turbulence flow like in the rotary or flash dryer. As algae cells can be damaged at 100°C, the temperature used in this type of drying method was lower (convective heat transfer mode at 80°C). This way, the risks of fire and emissions of volatile organic compounds (VOC) could be lowered. Further, lower temperature opened the option of recycling heat and using waste heat. The design was based on a production capacity of 1000 kg/h and aimed to reduce moisture content from 35–75% to 10%. The result showed that the cost to dry the microalgae from 55 to 10% ranged from $46.13 to $109.64 per tonne of dried product. It varied in accordance with the Hand factor (ratio of installed cost over the purchased cost). On the other hand, the drying cost was lower when the conveyor belt dryer used waste heat rather than natural gas like the commercial spray drier.

5.4.2 POLYPROPYLENE NON-WOVEN MEMBRANE DRYING

Recently, Sahoo et al. (2017) published a fresh work on polypropylene non-woven fabric membrane (PNM) dewatering and drying of Scenedesmus sp. and Chlorella sorokiniana. PNM was found to be an effective alternative for algae drying when supported by wire gauge (WG). The results showed that moisture was reduced to 10.5 and 7.2% (fully dried) in 2 h and 24 h, respectively, when left to dry on a PNM supported with a wire gauge stand and, subsequently, on a sand bed.

On another note, unlike freeze drying, spray drying, and oven drying, natural drying methods bring some advantages which include energy efficiency and cost effectiveness. Also, it is worth mentioning that the chemical properties of biomass may be influenced by the drying methods and parameters. Therefore, the authors investigated the drying conditions with intrinsic properties such as the influence on carbohydrate, protein, and lipid content in algae cakes. Evidently, there is no significant difference to algae properties when using three different natural drying methods, e.g. (1) slow wind, (2) shade drying ($27 \pm 2°C$, 20 km/h wind speed), and (3) sun drying (45–55°C floor temperature; $32 \pm 3°C$ ambient temperature) with 20 km/h average wind speed. Of these, sun drying showed the best result, as it dried almost three times fast and the moisture content was less than 20%. It is also argued that this drying method can be relatively cheaper and easier to apply; however, it is weather dependent and there will be a risk of algae fermentation; also algae cells may be damaged if drying time is prolonged. Moreover, it needs more space and time. In countries that have limited land for algae processing, it appears to be unsuitable for industry practice. In a nutshell, PNM serves as an emerging solution for dewatering and drying of algae as it is low cost, has re-use potential, and is effective. By tapping into the advantage of natural drying, it can be more environmentally friendly and economical.

5.5 CHALLENGES AND PROSPECTS

There are controversial arguments on the utilisation of algae in biofuel processing, and they are delineated well in Vassilev and Vassileva (2016). With regard to drying technology, the challenges and prospects to the nascent algae biofuel application are illustrated in Figure 5.5 and outlined here.

Firstly, in industry practice, cost effectiveness will be one of the main concerns (Chen et al., 2015). Studies suggested that the cost required for heating may constitute up to 50% of the overall operational cost. A comprehensive drying system or integrated drying system may be designed for optimum drying rate and energy consumption as well as prevention of degradation, which damages intrinsic properties. The design of a novel drying system may have its highlights in the recycling of heat and application of alternative green energy such as solar panels or wind energy to minimise the usage of fossil fuels. The adoption of a layered drying system through hot air or heat circulation or insulation in a chamber may be viable for countries that have limited land use. Further, it should be able to control the internal temperature to

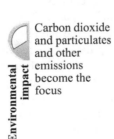

Industry practice	Environmental impact	Technology
Optimum drying rate, energy consumption and prevent degradation or damaging the intrinsic properties.	Carbon dioxide and particulates and other emissions become the focus	Membrane, extracellular polysaccharide substances (EPS) in microorganism and reducing or reutilizing raw material or byproduct

FIGURE 5.5 Challenges and prospects for algae drying.

achieve optimum drying conditions in order to produce the desired yield and properties in the subsequent process, e.g. transesterification, fermentation, etc. Also, the
design can take into consideration the gravitational effect for conveying or collecting
the dried matter. This can further reduce the operational cost for collection as well
as an additional conveyor. Show et al. (2015) noted that most of the algae drying
methods are developed from conventional dewatering of sewage sludge. Thus, it is
suggested that an innovative design of a belt dryer may be an option. This drying
method is not popular; however, it could dry up to 90% solid content with an electricity consumption of 400 Wh and 13.8 MJ of heat per kg of dry matter processed
(Lardon et al., 2009; Hassebrauck and Ermel, 1996).

Secondly, the environmental impact. Many have pointed out the utilisation of fossil fuels as a source for the drying of algae. The energy used is massive. In addition,
carbon dioxide and particulates and other emissions become another focus area.
Furthermore, in order to cultivate algae on a large industrial scale, arable land is utilised. Intangible costs may vary in different countries. Therefore, it is useful to utilise
effective tools to help to analyse the feasibility of a drying method. The solution to
this may be the application of life cycle assessment (LCA). LCA is a comprehensive
assessment method for the overall process which has been defined in goal and scope.
Studies have been conducted in utilising this tool for evaluations in other processes,
for instance, dewatering. The LCA results from Sander and Murthy (2010) showed
that one of the dewatering methods, thermal dewatering of algae, required a high
amount of fossil fuel-derived energy and presented a momentous reduction in energy
use. The study concluded that the use of algae as a fuel source was not feasible, and
a significant reduction in energy use should be pursued for future technology design
and to make algae-originated biofuel a sustainable and practical option as a fossil
fuel alternative. However, LCA results of Gnansounou and Kenthorai Raman (2016)
revealed that algae systems producing biodiesel, animal feed, and succinic acid were
beneficial to the environment as their renewability can reduce potential impacts.
Clearly, thermal drying caused an adverse impact, but this could be compensated
by the products. On another note, data input should be carefully conducted to avoid
differences and inconsistencies such as in system boundary definition, depreciation
method discount rate, assumptions used, and so on (Quinn and Davis, 2015; Quinn
et al., 2014). The result may serve as a reference for engineering approach to developing LCA model.

The preparation of membranes requires some chemicals that may raise costs
and cause negative environmental impacts. Selection of chemicals and materials
is needed during the preparation of the membrane product. As the utilisation of
material may be costlier, including the energy required for operation, assessment in
view of the materials and economic aspect can be performed using techno-economic
assessment (TEA). The outcome is postulated to greatly help the industry stake-
holder to anticipate its feasibility. When comparing between two drying methods
and multi-criteria to be evaluated, a systematic methodology, namely, fuzzy analytic
hierarchy process (FAHP) approach, can be employed (Jully Tan et al., 2016).

Green technologies emerge for sustainable growth. A promising solution is sought
to lead the future direction. Furthermore, innovative solutions always need low costs,
high efficiency, competent people to manage the system, and low environmental risk.

Often, physical treatment solutions require man power, chemical approaches pose environmental risks, and biological methods require competent people to manage them. Weighing the pros and cons, prominent solutions are suggested here for future research.

To achieve both high efficiency and lower environmental risk, an integration between chemical and biological methods could be applied. Culturing suitable microorganisms can help with dehydration and thus with the drying process. It is well known that extracellular polysaccharide substances (EPS) in microorganisms are capable to destabilise the microalgae. Tapping into this benefit, a dehydration process with osmotic pressure could be conducted to further enhance the drying of algae as indicated in Mazzuca Sobczuk et al. (2015).

For a low-cost approach, reducing or reutilising raw materials or byproduct is a feasible option to adopt. In view of green chemistry, renewable feedstock can be applied to microalgae cultivation, as well as glycerol (byproduct) utilisation as a draw solution in forward osmosis (Mazzuca Sobczuk et al., 2015). Reducing raw materials can be achieved by optimising the biofuel production in kg/area and minimising the chemicals used. Thus, chemical reactions and processes related to transesterification, fermentation, etc. processes should be further investigated.

5.6   CONCLUSION

Recognising the design, processes, and operating conditions of various drying methods at commercial practice level are crucial to weigh the disproportionate effects to the environment while addressing the limitations on energy use and cost incurred. More studies are needed to predict how to best account for drying efficiency despite all these considerations. Furthermore, researchers and industry stakeholders should explicitly consider the utilisation of natural sources of energy.

REFERENCES

Abomohra, Abd El-Fatah, Mostafa El-Sheekh, and Dieter Hanelt. 2014. "Pilot cultivation of the chlorophyte microalga Scenedesmus obliquus as a promising feedstock for biofuel." *Biomass and Bioenergy* 64:237–244. doi: http://dx.doi.org/10.1016/j.biombioe.2014.03.049.

Adams, J. M. M., A. B. Ross, K. Anastasakis, E. M. Hodgson, J. A. Gallagher, J. M. Jones, and I. S. Donnison. 2011. "Seasonal variation in the chemical composition of the bioenergy feedstock Laminaria digitata for thermochemical conversion." *Bioresource Technology* 102 (1):226–234. doi: http://dx.doi.org/10.1016/j.biortech.2010.06.152.

Aziz, Muhammad, Takuya Oda, and Takao Kashiwagi. 2013. "Enhanced high energy efficient steam drying of algae." *Applied Energy* 109:163–170. doi: http://dx.doi.org/10.1016/j.apenergy.2013.04.004.

Aziz, Muhammad, Takuya Oda, Takashi Mitani, Takumi Kurokawa, Norihiro Kawasaki, and Takao Kashiwagi. 2015. "Enhanced energy utilization system of algae: Integrated drying, gasification and combined cycle." *Energy Procedia* 75:906–911.

Balasubramanian, Sundar, James D. Allen, Akanksha Kanitkar, and Dorin Boldor. 2011. "Oil extraction from Scenedesmus obliquus using a continuous microwave system – design, optimization, and quality characterization." *Bioresource Technology* 102 (3):3396–3403. doi: http://dx.doi.org/10.1016/j.biortech.2010.09.119.

Barros, Ana I., Ana L. Gonçalves, Manuel Simões, and José C. M. Pires. 2015. "Harvesting techniques applied to microalgae: A review." *Renewable and Sustainable Energy Reviews* 41:1489–1500. doi: http://dx.doi.org/10.1016/j.rser.2014.09.037.

Bennion, Edward P., Daniel M. Ginosar, John Moses, Foster Agblevor, and Jason C. Quinn. 2015. "Lifecycle assessment of microalgae to biofuel: Comparison of thermochemical processing pathways." *Applied Energy* 154:1062–1071. doi: http://dx.doi.org/10.1016/j .apenergy.2014.12.009.

Brennan, Liam Philip Owende 2010. "Biofuels from microalgae—A review of technologies for production, processing, and extractions of biofuels and co-products." *Renewable and Sustainable Energy Reviews* 14 (2):557–577. doi: http://dx.doi.org/10.1016/j.rser.20 09.10.009.

Brink, Jacobus, and Sanette Marx. 2013. "Harvesting of Hartbeespoort Dam micro-algal biomass through sand filtration and solar drying." *Fuel* 106:67–71. doi: http://dx.doi.o rg/10.1016/j.fuel.2012.10.034.

Carrère, H., C. Dumas, A. Battimelli, D. J. Batstone, J. P. Delgenès, J. P. Steyer, and I. Ferrer. 2010. "Pretreatment methods to improve sludge anaerobic degradability: A review." *Journal of Hazardous Materials* 183 (1):1–15. doi: http://dx.doi.org/10.1016/j.jhazmat.2010.06.129.

Chen, Ching-Lung, Jo-Shu Chang, and Duu-Jong Lee. 2015. "Dewatering and drying methods for microalgae." *Drying Technology* 33 (4):443–454. doi: https://doi.org/10.1080/0 7373937.2014.997881.

Chen, Paris Honglay, and William J. Oswald. 1998. "Thermochemical treatment for algal fermentation." *Environment International* 24 (8):889–897. doi: http://dx.doi.org/10.1016/ S0160-4120(98)00080-4.

Chen, Wan-Ting, Junchao Ma, Yuanhui Zhang, Chao Gai, and Wanyi Qian. 2014. "Physical pretreatments of wastewater algae to reduce ash content and improve thermal decomposition characteristics." *Bioresource Technology* 169:816–820. doi: http://dx.doi.o rg/10.1016/j.biortech.2014.07.076.

Chheda, P., Grasso, D., van Oss, C.J. 1992. "Impact of ozone on stability of montmrillonite suspensions." *Journal of Colloid and Interface Science* 153:226–236.

Choi, Seung Phill, Minh Thu Nguyen, and Sang Jun Sim. 2010. "Enzymatic pretreatment of Chlamydomonas reinhardtii biomass for ethanol production." *Bioresource Technology* 101 (14):5330–5336. doi: http://dx.doi.org/10.1016/j.biortech.2010.02.026.

Desmorieux, Hélène, and Nadège Decaen. 2005. "Convective drying of spirulina in thin layer." *Journal of Food Engineering* 66 (4):497–503. doi: http://dx.doi.org/10.1016/j.jf oodeng.2004.04.021.

Dott, Ralf Andreas Genkinger, and Thomas Afjei. 2012. "System evaluation of combined solar & heat pump systems." *Energy Procedia* 30:562–570. doi: https://doi.org/10.1016/j. egypro.2012.11.066.

El Hage, Hicham, Amal Herez, Mohamad Ramadan, Hassan Bazzi, and Mahmoud Khaled. 2018. "An investigation on solar drying: A review with economic and environmental assessment." *Energy* 157:815–829. doi: https://doi.org/10.1016/j.energy.2018.05.197.

Eltawil, Mohamed A., Mostafa M. Azam, and Abdulrahman O. Alghannam. 2018. "Energy analysis of hybrid solar tunnel dryer with PV system and solar collector for drying mint (MenthaViridis)." *Journal of Cleaner Production* 181:352–364. doi: https://doi.org/10.1 016/j.jclepro.2018.01.229.

Esquivel, Beatriz Cordero, Domenico Voltolina Lobina, and Francisco Correa Sandoval. 1993. "The biochemical composition of two diatoms after different preservation techniques." *Comparative Biochemistry and Physiology Part B: Comparative Biochemistry* 105 (2):369–373. doi: http://dx.doi.org/10.1016/0305-0491(93)90243-X.

Gerde, Jose A., Linxing Yao, JunYi Lio, Zhiyou Wen, and Tong Wang. 2014. "Microalgae flocculation: Impact of flocculant type, algae species and cell concentration." *Algal Research* 3:30–35. doi: http://dx.doi.org/10.1016/j.algal.2013.11.015.

Gnansounou, Edgard, Jegannathan Kenthorai Raman. 2016. "Life cycle assessment of algae biodiesel and its co-products." *Applied Energy* 161:300–308. doi: http://dx.doi.org/10.1016/j.apenergy.2015.10.043.

González-Fernández, C., B. Sialve, N. Bernet, and J. P. Steyer. 2012. "Thermal pretreatment to improve methane production of Scenedesmus biomass." *Biomass and Bioenergy* 40:105–111. doi: http://dx.doi.org/10.1016/j.biombioe.2012.02.008.

Gregory, J. 1977. "Effect of polymers on colloid stability." Proc. of the NATO Adv. Study on the Scientific Basis of Flocculation, Cambridge, England, p. 1.

Guldhe, Abhishek, Bhaskar Singh, Ismail Rawat, Krishan Ramluckan, and Faizal Bux. 2014. "Efficacy of drying and cell disruption techniques on lipid recovery from microalgae for biodiesel production." *Fuel* 128:46–52. doi: http://dx.doi.org/10.1016/j.fuel.2014.02.059.

Gupta, Shilpi, Sabrina Cox, and Nissreen Abu-Ghannam. 2011. "Effect of different drying temperatures on the moisture and phytochemical constituents of edible Irish brown seaweed." *LWT – Food Science and Technology* 44 (5):1266–1272. doi: http://dx.doi.org/10.1016/j.lwt.2010.12.022.

Hassebrauck, Martin, and Gerrit Ermel. 1996. "Two examples of thermal drying of sewage sludge." *Water Science and Technology* 33 (12):235–242. doi: http://dx.doi.org/10.1016/0273-1223(96)00478-7.

Hosseinizand, Hasti, C. Jim Lim, Erin Webb, and Shahab Sokhansanj. 2017. "Economic analysis of drying microalgae Chlorella in a conveyor belt dryer with recycled heat from a power plant." *Applied Thermal Engineering* 124:525–532. doi: http://dx.doi.org/10.1016/j.applthermaleng.2017.06.047.

Hulatt, Chris J., Orsolya Berecz, Einar Skarstad Egeland, René H. Wijffels, and Viswanath Kiron. 2017. "Polar snow algae as a valuable source of lipids?" *Bioresource Technology* 235:338–347. doi: http://dx.doi.org/10.1016/j.biortech.2017.03.130.

Jully Tan, Kok Yuan Low, Nik Meriam Nik Sulaiman, Raymond R. Tan, Michael Angelo B. Promentilla. 2016. "Fuzzy analytical hierarchy process (AHP) for multi-criteria selection in drying and harvesting process of microalgae system." *Clean Techn Environ Policy* 18:2049–2063.

Kaliyan, Nalladurai, and R. Vance Morey. 2009. "Factors affecting strength and durability of densified biomass products." *Biomass and Bioenergy* 33 (3):337–359. doi: http://dx.doi.org/10.1016/j.biombioe.2008.08.005.

Kaushik, Prachi, Anushree Malik, and Satyawati Sharma. 2013. "Vermicomposting: An eco-friendly option for fermentation and dye decolourization waste disposal." *CLEAN – Soil, Air, Water* 41 (6):616–621. doi: http://dx.doi.org/10.1002/clen.201200248.

Laamanen, Corey A., Gregory M. Ross, and John A. Scott. 2016. "Flotation harvesting of microalgae." *Renewable and Sustainable Energy Reviews* 58:75–86. doi: http://dx.doi.org/10.1016/j.rser.2015.12.293.

Lam, Man Kee, and Keat Teong Lee. 2012. "Potential of using organic fertilizer to cultivate Chlorella vulgaris for biodiesel production." *Applied Energy* 94:303–308. doi: http://dx.doi.org/10.1016/j.apenergy.2012.01.075.

Lam, Man Kee, and Keat Teong Lee. 2014. "Cultivation of Chlorella vulgaris in a pilot-scale sequential-baffled column photobioreactor for biomass and biodiesel production." *Energy Conversion and Management* 88:399–410. doi: http://dx.doi.org/10.1016/j.enconman.2014.08.063.

Lardon, Laurent, Arnaud Hélias, Bruno Sialve, Jean-Philippe Steyer, and Olivier Bernard. 2009. "Life-cycle assessment of biodiesel production from microalgae." *Environmental Science & Technology* 43 (17):6475–6481. doi: 10.1021/es900705j.

Leach, G., G. Oliveira, and R. Morais. 1998. "Spray-drying of Dunaliella salina to produce a β-carotene rich powder." *Journal of Industrial Microbiology and Biotechnology* 20 (2):82–85. doi: http://dx.doi.org/10.1038/sj.jim.2900485.

Liapis, Athanasios, and Roberto Bruttini. 2014. "Freeze drying." In *Handbook of Industrial Drying, Fourth Edition*, 259–282. Boca Raton: CRC Press.

Lin, Jr-Lin, Chihpin Huang, and W. M. Wang. 2015. "Effect of cell integrity on algal destabilization by oxidation-assisted coagulation." *Separation and Purification Technology* 151:262–268. doi: http://dx.doi.org/10.1016/j.seppur.2015.07.064.

Mazzuca Sobczuk, T., M. J. Ibáñez González, E. Molina Grima, and Y. Chisti. 2015. "Forward osmosis with waste glycerol for concentrating microalgae slurries." *Algal Research* 8:168–173. doi: http://dx.doi.org/10.1016/j.algal.2015.02.008.

Milledge, John, Benjamin Smith, Philip Dyer, and Patricia Harvey. 2014. "Macroalgae-derived biofuel: A review of methods of energy extraction from seaweed biomass." *Energies* 7 (11):7194. doi: http://dx.doi.org/10.3390/en7117194.

Minowa, T., and S. Sawayama. 1999. "A novel microalgal system for energy production with nitrogen cycling." *Fuel* 78 (10):1213–1215. doi: http://dx.doi.org/10.1016/S0016-2361(99)00047-2.

Mohn, F.H., and C.J. Soeder. 1978. "Improved technologies for the harvesting and processing of microalgae and their impact on production costs." *Archiv fur Hydrobiologie, Beihefte Ergebnisse der Limnologie* 1:228–253.

Nindo, C. I., and J. Tang. 2007. "Refractance window dehydration technology: A novel contact drying method." *Drying Technology* 25 (1):37–48. doi: http://dx.doi.org/10.1080/07373930601152673.

Ommen, Torben, Wiebke Brix Markussen, and Brian Elmegaard. 2014. "Heat pumps in combined heat and power systems." *Energy* 76:989–1000. doi: https://doi.org/10.1016/j.energy.2014.09.016.

Plummer, J.D., Edzwald, J.K. 2002. "Effects of chlorine and ozone on algal cell properties and removal of algae by coagulation." *Journal of Water Supply: Research and Technology – AQUA* 51:307–318.

Poblete, Rodrigo, Ernesto Cortes, Juan Macchiavello, and José Bakit. 2018. "Factors influencing solar drying performance of the red algae Gracilaria chilensis." *Renewable Energy* 126:978–986. doi: https://doi.org/10.1016/j.renene.2018.04.042.

Prajapati, Sanjeev Kumar, Arghya Bhattacharya, Anushree Malik, and V. K. Vijay. 2015. "Pretreatment of algal biomass using fungal crude enzymes." *Algal Research* 8:8–14. doi: http://dx.doi.org/10.1016/j.algal.2014.12.011.

Prakash, J., B. Pushparaj, P. Carlozzi, G. Torzillo, E. Montaini, and R. Materassi. 1997. "Microalgal biomass drying by a simple solar device." *International Journal of Solar Energy* 18 (4):303–311.

Prakash, Om and AnilKumar 2014. "Solar greenhouse drying: A review." *Renewable and Sustainable Energy Reviews* 29:905–910. doi: https://doi.org/10.1016/j.rser.2013.08.084.

Quinn, Jason C., and Ryan Davis. 2015. "The potentials and challenges of algae based biofuels: A review of the techno-economic, life cycle, and resource assessment modeling." *Bioresource Technology* 184:444–452. doi: https://doi.org/10.1016/j.biortech.2014.10.075.

Quinn, Jason C., T. Gordon Smith, Cara Meghan Downes, and Casey Quinn. 2014. "Microalgae to biofuels lifecycle assessment—Multiple pathway evaluation." *Algal Research* 4:116–122. doi: http://dx.doi.org/10.1016/j.algal.2013.11.002.

Raven, H Peter, and B. George Johnson. 1992. *Biology*. 3rd ed. Mosby-Year Book, Maryland Heights, Missouri.

Ryu, Byung-Gon, Jungmin Kim, Jong-In Han, Kyochan Kim, Donghyun Kim, Bum-Kyoung Seo, Chang-Min Kang, and Ji-Won Yang. 2018. "Evaluation of an electro-flotation-oxidation process for harvesting bio-flocculated algal biomass and simultaneous treatment of residual pollutants in coke wastewater following an algal-bacterial process." *Algal Research* 31:497–505. doi: http://dx.doi.org/10.1016/j.algal.2017.06.012.

Sahoo, Narendra Kumar, Sanjay Kumar Gupta, Ismail Rawat, Faiz Ahmad Ansari, Poonam Singh, Satya Narayan Naik, and Faizal Bux. 2017. "Sustainable dewatering and drying of self-flocculating microalgae and study of cake properties." *Journal of Cleaner Production* 159:248–256. doi: http://dx.doi.org/10.1016/j.jclepro.2017.05.015.

Sander, Kyle, and Ganti S. Murthy. 2010. "Life cycle analysis of algae biodiesel." *The International Journal of Life Cycle Assessment* 15 (7):704–714. doi: http://dx.doi.org/10.1007/s11367-010-0194-1.

Sanyano, Naruetsawan, Pakamas Chetpattananondh and Sininart Chongkhong. 2013. "Coagulation–flocculation of marine Chlorella sp. for biodiesel production." *Bioresource Technology* 147:471–476. doi: http://dx.doi.org/10.1016/j.biortech.2013.08.080.

Shelef, G., A. Sukenik, and M. Green. 1984. "Microalgae harvesting and processing: A literature review". Technion Research and Development Foundation Ltd., Haifa (Israel).

Show, Kuan-Yeow, Duu-Jong Lee, and Jo-Shu Chang. 2013. "Algal biomass dehydration." *Bioresource Technology* 135:720–729. doi: http://dx.doi.org/10.1016/j.biortech.2012.08.021.

Show, Kuan-Yeow, Duu-Jong Lee, Joo-Hwa Tay, Tse-Min Lee, and Jo-Shu Chang. 2015. "Microalgal drying and cell disruption – Recent advances." *Bioresource Technology* 184:258–266. doi: http://dx.doi.org/10.1016/j.biortech.2014.10.139.

Singh, Pushpendra, Vipin Shrivastava, and Anil Kumar. 2018. "Recent developments in greenhouse solar drying: A review." *Renewable and Sustainable Energy Reviews* 82:3250–3262. doi: https://doi.org/10.1016/j.rser.2017.10.020.

Slager, Bart, Athanasios A. Sapounas Eldert J. van Henten, and Silke Hemming. 2014. "Modelling and evaluation of productivity and economic feasibility of a combined production of tomato and algae in Dutch greenhouses." *Biosystems Engineering* 122:149–162. doi: https://doi.org/10.1016/j.biosystemseng.2014.04.008.

Syed, I.R., S. Sukhcharn, and D.C. Saxena. 2016. "Evaluation of physical and compositional properties of horse-chestnut (Aesculus indica) seed." *Journal of Food Processing and Technology* 7 (3):561.

Tenney, M.W., Echelberger, W.F., Schuessler, R.G., Pavpni, J.L. 1969. "Algal flocculation with synthetic organic polyelectrolytes." *Applied Microbiology* 18:965–971.

Vandamme, Dries. 2013. "Flocculation based harvesting processes for microalgae biomass production". Ph.D., Arenberg Doctoral School. Faculty of Bioscience Engineering, KU Leuven, Belgium.

Vandamme, Dries, Sandra Cláudia Vieira Pontes, Koen Goiris, Imogen Foubert, Luc Jozef Jan Pinoy, and Koenraad Muylaert. 2011. "Evaluation of electro – coagulation – flocculation for harvesting marine and freshwater microalgae." *Biotechnology and Bioengineering* 108 (10):2320–2329. doi: http://dx.doi.org/10.1002/bit.23199.

Vassilev, Stanislav V., and Christina G. Vassileva. 2016. "Composition, properties and challenges of algae biomass for biofuel application: An overview." *Fuel* 181:1–33. doi: http://dx.doi.org/10.1016/j.fuel.2016.04.106.

Venteris, Erik R., Richard L. Skaggs, Mark S. Wigmosta, and Andre M. Coleman. 2014. "A national-scale comparison of resource and nutrient demands for algae-based biofuel production by lipid extraction and hydrothermal liquefaction." *Biomass and Bioenergy* 64:276–290. doi: http://dx.doi.org/10.1016/j.biombioe.2014.02.001.

Villagracia, Al Rey C., Andres Philip Mayol Aristotle T. Ubando, Jose Bienvenido Manuel M. Biona, Nelson B. Arboleda, Melanie Y. David, Roy B. Tumlos, Henry Lee, Ong Hui Lin, Rafael A. Espiritu, Alvin B. Culaba, and Hideaki Kasai. 2016. "Microwave drying characteristics of microalgae (Chlorella vulgaris) for biofuel production." *Clean Technologies and Environmental Policy* 18 (8):2441–2451. doi: 10.1007/s10098-016-1169-0.

Wahlen, Bradley D., Mohammad S. Roni, Kara G. Cafferty, Lynn M. Wendt Tyler L. Westover, Dan M. Stevens, and Deborah T. Newby. 2017. "Managing variability in algal biomass production through drying and stabilization of feedstock blends." *Algal Research* 24 (Part A):9–18. doi: http://dx.doi.org/10.1016/j.algal.2017.03.005.

Widjaja, Arief, Chao-Chang Chien, and Yi-Hsu Ju. 2009. "Study of increasing lipid production from fresh water microalgae Chlorella vulgaris." *Journal of the Taiwan Institute of Chemical Engineers* 40 (1):13–20. doi: http://dx.doi.org/10.1016/j.jtice.2008.07.007.

Xu, Lixian, Derk W. F. Brilman, Jan A. M. Withag Gerrit Brem, and Sascha Kersten. 2011. "Assessment of a dry and a wet route for the production of biofuels from microalgae: Energy balance analysis." *Bioresource Technology* 102 (8):5113–5122. doi: http://dx.doi.org/10.1016/j.biortech.2011.01.066.

Zhang, X., Rong, J.F., Chen, H., He, C.L., Wang, Q. 2014. "Current status and outlook in the application of microalgae in biodiesel production and environmental protection." *Frontiers in Energy Research* 2:1–15.

Zheng, Hongli, Zhen Gao, Jilong Yin, Xiaohong Tang, Xiaojun Ji, and He Huang. 2012. "Harvesting of microalgae by flocculation with poly (γ-glutamic acid)." *Bioresource Technology* 112:212–220. doi: http://dx.doi.org/10.1016/j.biortech.2012.02.086.

6 Coal Drying in Large Scale
Simulation and Economic Analysis

Sankar Bhattacharya

CONTENTS

6.1 LOW-RANK COALS AND THE NEED FOR DRYING

Low-rank coals with a high-moisture (30–70% on an as-received weight basis) content represent a significant resource worldwide and generally consist of sub-bituminous coals and lignites. An estimated 45% of the world's coal reserves are lignites, some of which are referred to as brown coal. Most brown coals are cheap and low in ash and sulphur contents, but have high-moisture contents up to 70% on an as-received basis. Low-rank coals represent the major source for power generation in several countries – Australia (~50%), Germany (~75%), Greece (~90%), Poland

(~55%), Russia (~45%), the USA (~10%) are examples. Figure 6.1 identifies the major countries that use high ash or high-moisture or low-ash low-rank coals, and indicative ranges of moisture content, ash content, and calorific values of lignites in those countries.

Coal drying is an important and necessary step towards improving the efficiency of its utilisation – be it for gasification for power or fuel gas or high-value chemicals production, or for use in existing and new pulverised coal-fired power plants using high-moisture coals. In general, the efficiency of a power generation unit using coal drops by about 4% point and 9% point when coal moisture content increases from 10 to 40 and 60% respectively. This is quite significant as a 1%-point increase in efficiency can often result in up to 2.5%-point reduction in CO_2 emission.

Apart from efficiency reduction, high moisture in coal increases the requirement of auxiliary power in coal handling systems and pulverisers and therefore plant operating and maintenance costs. High moisture also makes the feeding of the moist coal into pressurised systems intermittent unless solids loading is reduced. However, drying high-moisture coals increases the risk of spontaneous combustion as, due to their high oxygen content, they are usually more reactive than higher-rank coals. Therefore, in most cases, drying plants have to be located close to the point of actual utilisation, be it combustion or gasification or other plants. This requirement poses a challenge for existing coal-fired units for space or their suitability from a heat transfer point of view.

While drying is a necessary and important step before the utilisation of high moisture, its adoption in a coal-fired plant depends on a range of factors which include the type of plant and size, whether it is new or retrofit, the initial moisture content in the coal, whether CO_2 capture is contemplated, and finally the cost. Therefore, a thorough evaluation of the techno-economics is vitally important.

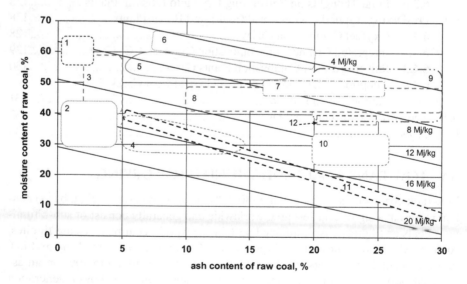

FIGURE 6.1 High-ash and/or high-moisture containing coals often termed as lignites – their locations and calorific values.

Despite their importance, drying plants have been implemented in large scale in only a handful of operating commercial coal-fired units. This reflects the complexities associated with their engineering implementation and their cost.

This chapter presents a summary of a handful of publically available studies and identifies the technical challenges and the immediate needs for research and development. Modelling of coal drying *per se* is not discussed in this paper. Also, in this chapter, we will largely focus on low-ash, high-moisture, low-rank coals.

6.2   MAJOR TECHNOLOGIES FOR LARGE-SCALE COAL DRYING

Advances in coal drying have been covered in Chapter 7, and therefore, in this chapter, we will only briefly introduce some of the frontrunner technologies in the context of the techno-economics of the use of drying at a large scale – in operating or new plants. As indicated in Chapter 7, drying can be categorised into evaporative and non-evaporative processes.

The non-evaporative methods release the water in coal as liquid water, and the process is, therefore, called dewatering. Evaporative methods release the water in the form of water vapour and therefore require energy for latent heat in the moisture to be evaporated.

Further broad classifications of the drying processes can be made based on the following:

*   **The velocity of the drying medium** – depending on the velocity, dryer types can be of the fluidised bed variety or entrained flow dryers. Fluidised bed dryers can tolerate a range of coarse and small coal particles (0–6 mm) for a longer residence time, whereas entrained flow (or flash) dryers preferentially need fine particles (<~200 μm) to dry the coal to the required degree within a reasonable residence time of seconds. Examples of the fluidised bed dryer include the WTA (Wirbelschicht Trocknung Anlage) dryer of the RWE, which has been implemented in Frechen and Nierderaussem plants for the drying of both coarse grain (0–6 mm, WTA1) and fine grain (0–2 mm, WTA2) lignites.

    Examples of the entrained flow dryer include COMBdry (Lee, 2018), under development by the Korean Institute of Energy Research (KIER) along with SMK Energy in Korea. This technology has been demonstrated for a range of lignites and biomass at 1 tonne/day scale with further demonstration slated for 2019 at 5 tonne/day scale.
*   **Drying medium** (such as steam, hot air) – WTA dryers use steam as drying medium, while DryFining™, the fluidised bed dryer developed by Great River Energy and implemented at the Coal Creek power station, is fluidised by hot air which in turn is heated with waste flue gas. More about the use of Dryfining™ technology is discussed in Section 6.4.4.
*   **Mode of contact between the medium and the coal particles** – the mode of contact between the coal and the drying medium can be direct or indirect. These are mainly tubular dryers or rotary drum dryers. Usually, indirectly heated dryers are used for very wet solids or thin product layer. Capital and

operating costs in an indirect dryer are usually higher compared to directly
heated dryers, but emissions and fire hazard from dried coals are lower.
Steam fluidised beds are an example of a mixed heating mode – where flu-
idisation of the coal being dried by steam is direct, but the major energy for
drying is supplied indirectly by slightly superheated steam.

- **Operating pressure** – usually all types of dryers operate at atmospheric
  pressure on the coal side, while the drying medium can be at above atmospheric pressure. Pressurised steam fluidised bed drying from 0.5 t/h to 10
  t/h scale where lignite at 55–60% moisture was dried 1–6 bar to a moisture
  content 8–17%.

  Pressurised steam fluidised bed drying has the advantage of higher heat
  transfer coefficient between the heating medium and the coal, resulting in a
  compact dryer. However, pressurised drying also means the need for feeding moist coal into a pressurised reactor, which has its share of difficulty.
  Development of pressurised steam fluidised bed drying appears to have
  stalled.

The major dewatering developments that took place in the past include hydrothermal dewatering (HTD) and mechanical thermal expression (MTE) in Australia and
Germany. These technologies are currently not frontrunners. Therefore, their use for
retrofit or greenfield application is not pursued at present.

6.3   RETROFIT OF DRYING TO EXISTING
COAL-FIRED BOILER PLANTS

Retrofitting drying plants to existing power generation plants improve the efficiency
of operation and hence reduce emissions per unit of electricity production. The extent
to which retrofit is possible depends on the type and design of the existing boiler, as it
is required to ensure heat flux matching inside the boilers, availability of space, and
compatibility with the existing auxiliaries system, in particular the coal handling
system. In general, purely from the boiler's point of view, up to 30% of coal input
can be replaced in pulverised lignite-fired boilers by pre-dried coals, but somewhat
more in lignite-fired circulating fluidised bed boilers due to their inherent flexibility
to handle different types of fuels.

6.3.1   RETROFITTING OF DRYING TO PULVERISED COAL-FIRED
UNITS AND CO-FIRING WITH EXISTING FUELS

In any case, retrofitting drying to existing units requires a case-by-case detailed heat
transfer and techno-economic analysis of operation during part and full loads. This
is often proprietary information, and as a result, very few studies are reported in
the published literature. In the following sections, we summarise major conclusions
from selected published studies.

In one of the early studies, Kakaras et al. (2002) presented simulation results
for the integration of an external dryer into a Greek lignite-fired power plant. Two

different types of dryers (a steam tubular dryer of indirect heating type and a WTA dryer) and one MTE dewatering system were considered. These systems with associated mass and heat balance data are shown in Figure 6.2a–c. The plant considered was a lignite-fired existing pulverised coal power unit in Greece of 365 MWe capacity where lignite was dried in-situ using flue gas. The lignite had 50–60% moisture content, and the two drying methods considered drying to 15% moisture content; in the case of the MTE dewatering, moisture content following dewatering was considered at 22%. The major objective of these simulations was to formulate the most efficient drying scheme which might be integrated into the steam cycle of the plant. For the drying systems, three different pressures were considered – 223 kPa, 518 kPa and 2 MPa steam pressure for the tubular dryer; 3.2 bar and 5 bar steam pressure for the WTA dryers. The results (in Figure 6.3) indicated that the highest efficiency increase was 7.39% for the MTE dryer, followed by 6.9% for the WTA dryer operating with 3.2 bar steam pressure. No economic analysis was carried out as part of this study.

Atsonios et al. (2016) extended the simulation work by Kakaras et al. (2002) with a WTA dryer fitted to the same plant in Greece but taking into account a more realistic approach of the load fluctuation during the day. They examined the effect of three concepts of drying. The same lignite was considered, but replacing 25% of the thermal load of the existing fuel.

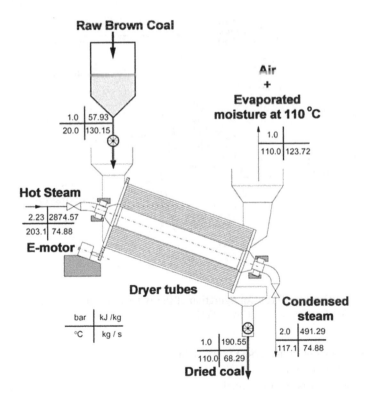

FIGURE 6.2 (a) Configuration of the steam-heated indirect tubular dryer and the data used (Kakaras et al., 2002).

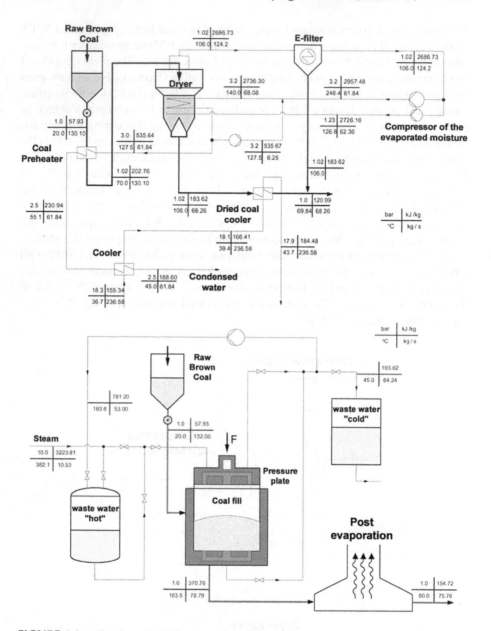

FIGURE 6.2 (Continued) (b) Configuration of the WTA dryer and the data used (Kakaras et al., 2002). (c) Configuration of the MTE dewatering system and the data used (Kakaras et al., 2002).

The first concept considers (Figure 6.4a) continuous operation of the dryer unit. In the second concept (Figure 6.4b), the drying plant operates intermittently. When the power demand is medium during the day, the plant operates following the first concept. During the night, the power plant operates at a higher load than the demand from the grid and the excess power/heat is utilised for the pre-drying of the lignite

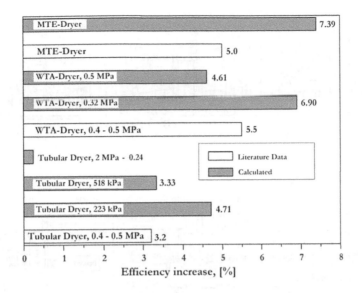

FIGURE 6.3 Comparison of the calculated net increase of efficiencies using different types of dryers (Kakaras et al., 2002).

for utilisation the following evening. During the evening, the plant operates at full load as the power demand is high, achieving the lowest electricity cost, apart from raw lignite, the pre-dried one, which had been produced and stored from the previous night.

The third concept (Figure 6.4c) considers drying plant operation only during the night. A larger dryer and hence higher investment costs are required in this concept, but the dryer operational cost is lower as it runs only for six hours as opposed to 24 hours.

The major conclusion is that flexible operation of lignite power plants can be carried out in a profitable way using the pre-drying technology. The flexible operation of the pre-drying system in certain periods during the 24 hours, taking advantage of the fluctuations of the electricity price for profit maximisation, "is a beneficial energy storage option." The cost of the dryer is a significant determinant of the best operational scheme. The average electricity cost in all three concepts varied from €32.55 to €33.39/MWh.

Drosatos et al. (2017) carried out computational fluid dynamics (CFD)-based simulation of the main furnace and the convective section of the same unit in Greece (Kakaras et al., 2002) at different thermal loads of 100, 60, and 35%. The objective was to ascertain the char conversion, gas distribution, temperature, and heat flux profile when pre-dried (12% moisture) lignite was partially introduced into the boiler where dried lignite (having 20% moisture) were combusted. The CFD results indicated that the introduction of pre-dried lignite facilitated stable combustion at low load (35%), which is lower than the manufacturer's suggested minimum of 55% of the nominal load. Char conversion was acceptably high for all loads, but the NOx concentration increased for lower loads.

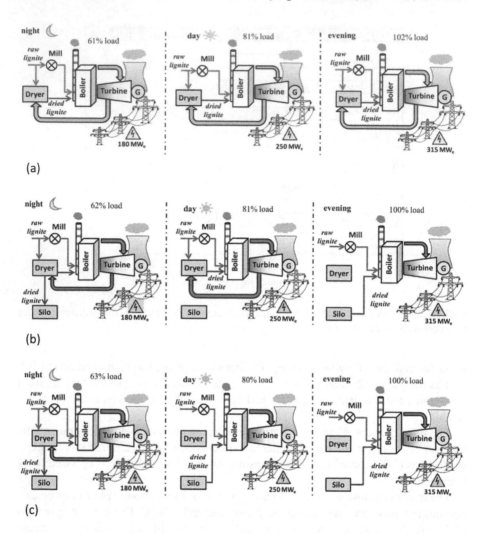

FIGURE 6.4 Schematic representation of three concepts of drying plant operation , (a) continuous drying, (b) drying during the night and the day, and (c) drying during the night using a larger dryer (Atsonios, 2016).

Han et al. (2017) investigated the suitability of water extraction from high-moisture lignite using efficient integration of waste heat and water recovery technologies with a flue gas pre-drying system. Two options for water extraction were considered in the flue gas pre-dried lignite-fired power system (FPLPS). The FPLPS integrates the fan mill flue gas dryer with an open pulverising system; this process increases the boiler efficiency. The dryer exhaust gas contains a large amount of vapour during the drying of high-moisture lignite, thereby there is a potential for waste heat and water recovery. The two options include the low-pressure economiser (LPE) for water-cooled units, and the spray tower (SPT) integrated with a heat pump for air-cooled units. Both options were found to have payback periods of around three years. The schematics of the two systems are reproduced in Figure 6.5a,b.

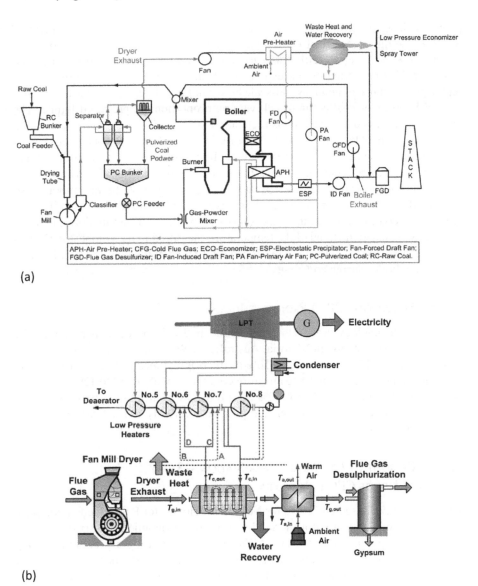

FIGURE 6.5 Schematic of the (a) FPLPS and (b) LPE systems (Han et al., 2017).

The FPLPS scheme indicated that the improvement in the efficiency of the water-cooled unit was larger than that of the air-cooled unit. The efficiency benefits depended on the initial moisture content of the lignite. The improvement was reduced from 1.47 to 1.20% in water-cooled units and from 1.40 to 1.14% in air-cooled units, when the moisture content of raw lignite decreased from 39.5 to 29.95%, and that of the pre-dried lignite kept constant as 9.82%. The water recovery ratio of the SPT scheme resulted in 0.31 t/MWh water-saving and 110.6 MWtth district heating supply by the integration of a heat pump with COP of 5.0. For the FPLPS system, the plant efficiency improvement was up to 1.47% for an initial moisture content of 39.5% in the lignite. For the LPE scheme, this improvement was around 0.20%.

Pawlak-Kruczek et al. (2018), as part of the DRYLIG project supported by the European Union, carried out an experimental and mathematical modelling (CFD) on the performance of an industrial-scale boiler in Poland during combustion of pulverised bituminous coal co-fired with various proportions of pre-dried lignite. The bituminous coal had an 8.55% moisture content and 15.86% ash content. The initial moisture content of the lignite was 41.73% dried to 26.51% and ash content of 12.57%.

Experiments indicated that co-milling of the bituminous coal and both wet and dried lignite was possible and, in fact, improved the grindability. Experiments also indicated that incorporating dried lignite during co-firing of the bituminous coal reduced the NOx emission, not unexpected as the lignite had less nitrogen than the bituminous coal; this is also the trend suggested by the CFD simulation. The CFD modelling matched the trend of the temperature measured in the combustor. On this basis, further CFD simulations were carried out to predict the flow and heat flux patterns in the boiler during co-firing.

6.3.2 HAZELWOOD 2030 PROJECT

The WTA technology was also considered in 2007 for demonstration at the Hazelwood power station in Victoria, Australia. A WTA dryer was to be retrofitted to an existing 200 MWe unit to dry 50% of the original feed of high-moisture coal, with the intention of reducing the moisture content from about 60 to 12%. The dried coal would then be co-fired with 50% high-moisture coal into the boiler (Burnard and Bhattacharya, 2011). However, the project eventually did not proceed due to unknown reasons.

6.3.3 RETROFIT TO LOY YANG POWER STATION – STUDY BY LIGNITE CRC IN VICTORIA

The Cooperative Research Centre for New Technologies for Power Generation from Low-Rank Coal carried out a thermodynamic and economic analysis of integrating coal drying plants into the 500 MWe units in Loy Yang power station using Victorian brown coal. A similar analysis was also carried out for two other power stations located in the same region but using different types of brown coals. Two different types of dryers were considered – steam fluidised bed dryer (SFBD), and dewatering technology – MTE. The prevalent drying technology used in the plant to dry 60% moisture-containing brown coal is mill drying using flue gas; this drying method is thermally inefficient as it consumes over 25% of the available energy in the coal. A proprietary boiler performance model was developed. The model predictions suggested that a reduction in the moisture content of the pre-dried coal would have a significant effect on the boiler efficiency, gas flows, coal consumption, and demand on the coal milling plant.

The furnace exit temperature was predicted to rise by 65°C if the water content in the pre-dried coal was reduced by half from 60%. There would be a redistribution of heat transfer between the economiser, waterwalls, superheater, reheater, and air heater, which was found to change with a reduction in the moisture content of the pre-dried coal. There would be a reduction in heat transfer to the economiser sections and consequently an increase in the furnace heat transfer which could potentially increase the slagging in the furnace region of the boiler. The redistribution of the heat transfer within the boiler could be compensated by either reduction in spray water flow or flue gas recirculation or applying excess air. An economic evaluation of the proposed integrated coal drying plants showed that none of the proposed methods was economically viable (as of 1998) in the absence of a CO_2 tax in the range of $25–$49/tonne, given the very low cost of the coal.

6.3.4 DryFining™ Retrofit to Loy Yang Power Station in Victoria

The Global CCS Institute (accessed August 2018), published a report on the assessment of retrofitting post-combustion capture to the Loy Yang Power A (LYA) in Victoria. As part of that study, an assessment was made on the applicability and efficiency improvements possible by pre-drying using the DryFining™ technology developed by Great River Energy.

At 60 wt%, Loy Yang's brown coal is also significantly higher in moisture content compared with Great River Energy's 38 wt% North Dakota lignite. Reducing the moisture in Loy Yang's coal feed has, therefore, the potential to provide a significant improvement in power station performance. In that study, the size of the coal drying plant was based on drying coal for one complete 500 MW plant (Unit 3) at LYA. The major source of waste heat for DryFining™ is the flue gas system. This heat is extracted by locating flue gas coolers (FGCs) in the flue gas stream at a convenient location downstream of the air heaters and particulate removal equipment. LYA operates flue gas at temperatures typically between 170 and 190°C at the ID fan inlets. However, the operating philosophy limit is not lower than 160°C, which is just 20–25°C above the sulphur dew point of that of the stack gas.

Even though corrosion due to sulphuric acid condensation in the flue gas system has been a serious problem in the LYA operating history, that study assessed the potential available at a stack temperature on the estimated dew point with a margin of 5°C above that point. Allowing for a design margin of 5°C, a lower stack gas temperature of 140°C may be used. This temperature equates the exit temperature of the flue gas cooler. The resulting heat available for coal drying then suggested that a fuel moisture reduction from 60 to 54% is possible.

Reduction of the moisture content to 54% results in a reduction of coal flow to the boiler and through the majority of the coal handling system. The reduced moisture load on the boiler results in an increased boiler efficiency from 72 to 75.4%.

6.3.5 Use of WTA Dryer in German Power Stations

The WTA system (RWE.com, 2018) was developed by RWE Rheinbraun in Germany. WTA uses recycled steam from coal moisture to fluidise the wet lignite in the dryer. The energy for drying comes mainly from the steam bled from the low-pressure turbine via a heat exchanger tubes immersed in the fluidised bed.

In 2007, a commercial prototype WTA dryer was commissioned in the supercritical Unit K of RWE's Niederaussem power plant (BoA concept with pre-dried lignite). This dryer can reduce the lignite moisture from 50–55 to approximately 12%. Pre-dried lignite can form up to 30% of the fuel on an energy basis, which can lead to a 1%-point gain in the plant thermal efficiency and a 2.5% reduction in CO_2 emissions. In 2008, RWE reported a total investment of € 50 million for the erection and operation of this WTA dryer.

For a supercritical lignite-fired power plant firing 100% pre-dried lignite (BoA Plus concept of RWE), RWE considered a specific investment cost of €70/kW for the open cycle variant of WTA, which included costs of modifying the burners and boiler combustion chambers. The overall investment costs of a BoA Plus are similar to those of a BoA unit, but the thermal efficiency could increase by 4–5% point.

6.3.6 Final Thoughts on Retrofitting Drying to Existing Plants

Techno-economic information on modern pre-drying processes and retrofitting to existing plants is scarce and incomplete in the public literature. Most of the published information is based on process or CFD simulation. It is difficult to experimentally verify the CFD predictions at full scale, but often the modelling methodology is developed by validating experimental results from small or pilot-scale rigs. Nevertheless, such predictions for large-scale applications are used as first-cut considerations before full-scale implementation can be considered. This has to be carried out on a case-by-case basis.

6.4 PRE-DRYING APPLIED TO GREENFIELD COAL-FIRED BOILER PLANTS

Any new high-moisture lignite-fired plant, combustion, or gasification, will require pre-drying the coal. It is easier to build drying plants for new or greenfield sites as there are fewer design constraints due to a lack of appropriate space or limited flexibility with boiler design.

In the subsequent sections, we present plant design concepts, thermodynamic process simulations, and techno-economic considerations used in different studies for building new plants with lignite pre-drying for combustion and gasification plants. Among the combustion plants, we consider all three major types – currently used or under consideration – the pulverised coal type, the circulating fluidised bed boiler types which are commercially available, and the oxyfuel boiler which was under development. While several studies are available in published literature, the results of a selection of a few are presented here.

6.4.1 Oxyfuel Combustion Cases

In one of the early studies as part of the ENCAP project supported by the European Union, Kakaras et al. (2007) examined a greenfield application of the oxyfuel technology for CO_2 sequestration in a lignite-fired (55% initial moisture) supercritical power plant of 360 MWe net capacity. The lignite was assumed to be pre-dried to

12% moisture (close to equilibrium moisture) using a WTA dryer operating at 106°C drying temperature. Oxyfuel technology requires an air separation plant which was considered to be of a cryogenic type. In a greenfield design, it is possible to include measures for the reuse of waste heat, and therefore heat integration from processes such as air separation (ASU), CO_2 compression, purification, and flue gas treatment were used to reduce the energy and efficiency penalties. In such schemes, different heat integration options are possible and result from thermodynamic simulations dealing with the most important features of the oxyfuel power plant with CO_2 capture. A comparison with a retrofit application is also presented. This study did not include economic considerations. For the optimised configuration of the heat integration options with a CO_2 capture efficiency of 90%, about 35% of the net auxiliary power consumption was due to the ASU while the drying plant used about 17%. The gross power output increased by about 10% due to the adoption of pre-drying.

Cormos (2016) presented a detailed techno-economic analysis for an oxyfuel power plant to generate about 350 MWe net power with a CO_2 capture rate of at least 90%. Romanian lignite with 40% moisture was used and pre-dried in a WTA dryer with vapour compressor to fuel a supercritical power plant (main steam pressure 582°C/29 MPa). The plant simulation was carried out using ChemCAD and Thermoflow software. For the optimised configuration of the heat integration options with an MDEA-based CO_2 capture efficiency of 93%, about 37% of the net auxiliary consumption was due to the ASU while the drying plant used about 15%. The cost of the WTA plant was about 4.25% of the total capital cost of €2671/kWe, net which included the cost of ASU, flue gas desulphurisation, CO_2 processing, and conditioning, all of which have higher capital cost than the WTA dryer.

Kotowicz and Balici (2014) simulated a supercritical oxyfuel circulating fluidised bed boiler fed with lignite containing 42.5% moisture. An ASU unit based on a three-end type high-temperature membrane and CO_2 compression installation was used. The waste air from the ASU was used for drying the lignite, which, according to the authors, would be mimicking a fluidised bed dryer operating at 130°C and drying the lignite under 20%. The steam cycle operated at 600°C/290 bar and 620°C/50 bar for main steam and reheat temperature respectively generating 600 MW gross power. The incorporation dryer drying the lignite to 10% moisture content increased the thermal efficiency by 3.9% point.

6.4.2 INTEGRATED GASIFICATION COMBUSTION CYCLE (IGCC) CASE

Fushimi and Dewi (2015) simulated an integrated gasification combined cycle (IGCC) plant fired with sub-bituminous coals and lignites having 26 and 36% moisture content respectively. The lignite was assumed to be dried in an indirectly heated dryer with superheated steam with facilities for self-heat recuperation (SHR) to 12% moisture content. SHR-based drying is claimed to be more energetically efficient than conventional mechanical vapour compression (MVR)-based superheated steam dryers. The gasification plant was assumed to be of Shell type with an operating temperature of 1371°C for lignite. The fuel gas was utilised in a gas turbine at 1500°C turbine inlet temperature. The net thermal efficiency of the lignite plant was estimated to be 41.5% before drying. The estimated fixed capital investment for the drying plant is 1.86–4.86% of the total capital cost of the IGCC plant which was US$1.544 billion for 280 MWe net output. The lignite input to the drying plant was 95.76 kg/s. The fixed cost of the drying plant equates to US$ 29–75 million. The efficiency drop due to the auxiliaries, including the ASU, coal crushing, and pulverising is about 4.5% point.

6.4.3　PULVERISED COAL COMBUSTION PLANTS CASES

Xu et al. (2014) conducted a techno-economic analysis for a 600 MWe supercritical pulverised lignite-fired plant fitted with a rotary steam dryer drying the lignite from 39.5 to 15% moisture. Supercritical steam at 5 bar, 290°C was extracted from the low-pressure turbine for drying conducted at 1.5 bar. While net electrical efficiency increased by 2.6% as a result of pre-drying, the cost of electricity (COE) decreased by A$1.2/MWh. The full details of the fixed capital costs for the entire plant and the drying plant were not available.

Zhu et al. (2018) simulated a pulverised lignite-fuelled power plant of 600 MWe capacity integrated with a two-stage drying system which consisted of two fluidised bed dryers and an additional feed water heater, as shown in Figure 6.6. The first stage dryer used flue gas from the boiler both for fluidising and as a drying heat source. The second stage dryer was a steam fluidised bed dryer. The initial moisture content of the lignite was 36.2%, that is 0.362 kg/kg of lignite. The thermodynamic and economic analysis showed that at lignite drying degrees of 0.1, 0.2, and 0.3 kg/kg, the power generation efficiency of the power plant is 1.45, 2.12, and 2.81% higher than that of the conventional lignite power plant without drying. The annual net economic benefit was estimated to be 1.34, 2.03, and 1.60 M$ during the lifetime of the drying system.

The Zero Emissions Platform (ZEP, 2011) sponsored by the European Union carried out a comprehensive study examining the complete costs of CO_2 capture and

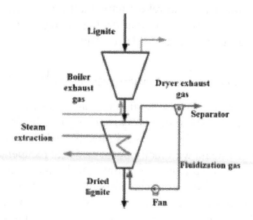

FIGURE 6.6　Configuration of the two-stage dryer (Zhu et al., 2018).

storage (CCS) for newly built power plants fired with hard coal, lignite, and natural gas. Three capture technologies were considered – post-combustion, pre-combustion, and oxyfuel technology, and the levelised cost of electricity (LCoE) was calculated along with CO_2 avoidance costs as of 2009.

For the lignite, three different pulverised fuel cases were calculated and compared:

- with state-of-the-art ultra-supercritical (USC) steam parameters ignite-fired 989 MWe net PF ultra-supercritical (280 bar 600/620°C steam cycle) power plant without capture and pre-drying;
- a lignite-fired 920 MWe net PF ultra-supercritical (280 bar 600/620°C steam cycle) power plant with pre-drying of the lignite but without capture;
- a lignite-fired 1100 MWe net PF ultra-supercritical (280 bar 600/620°C steam cycle) power plant with pre-drying of the lignite but without capture.

The lower heating value-based plant efficiency was estimated to be 43%, 49%, and 48% respectively for the three cases. Pre-drying lowered the levelised fuel costs from €13.6/MWh to €12/MWh. It is not prudent to calculate any LCoE in these cases to ascertain the effect of pre-drying as there are interrelated factors dependent on both fixed capital and fixed O&M costs.

6.4.4 Adoption of DryFining™ Technology at Coal Creek Power Station, North Dakota

Great River Energy (GRE) has developed and patented a coal drying and coal upgrading technology termed DryFining™. This technology improves the efficiency of a coal-fired power station and thereby reduces CO_2 emissions. The DryFining™ concept involves utilising waste heat, which may be available in a power station, to partially dry the feed (raw) coal in a fluidised bed dryer (FBD). Using the gravitational segregation feature of a fluidised bed and the special design of a coal dryer, denser materials present in the coal such as pyrites, small rocks, and sands are segregated and separated from coal, thereby improving the quality of the coal fed to the power station. DryFining™ was developed in conjunction with the first round of the U.S. Department of Energy (DOE) Clean Coal Power Initiative (CCPI) in 2002 and with the additional support of the National Energy Technology Laboratory (NETL). GRE proved the technical feasibility of the concept in three stages. Firstly in a proof-of-concept 2 t/h pilot-scale coal dryer, then a 75 t/h prototype-scale dryer integrated in the GRE Coal Creek Station, located in North Dakota, USA, and in the third stage in the complete conversion of this 2×600 MW mine-mouth power station to DryFining™. The technology is in commercial scale operation in both Unit 1 and Unit 2 at GRE's Coal Creek plant. Coal moisture can be reduced from 3% to 29%. There are fuel savings of about 4% and a 4% gain in the overall plant thermal efficiency.

The schematic of the integrated DryFining™ dryer at the Coal Creek power station is shown in Figure 6.7 (National Energy Technology Laboratory, 2011). The commercial coal drying system at Coal Creek includes four commercial sizes (125 t/h) moving bed fluidised bed dryers per unit, crushers, a conveying

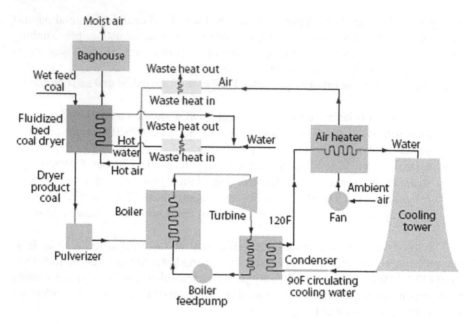

FIGURE 6.7 Schematic of the integrated DryFinding™ dryer at the Coal Creek power station (Burnard and Bhattacharya, 2011).

system to handle raw lignite, segregated, product streams, particulate control system, and control system. The system commissioning was completed in December 2009. Reportedly the first commercial installation incurred investment costs of US$ 240/kWnet, but these costs could be reduced toUS$80–100/kWnet for future retrofit installations. The O&M costs for the dryers at Coal Creek station were estimated at US$0.35/wet tonne of lignite processed or US$ 350 000/year for a 125 short tph dryer. But pre-drying of coal in both units resulted in more than US$20 million/year savings in fuel, auxiliary power consumption, and other O&M costs (World Coal, 2014).

6.5 THE WAY FORWARD ON COAL DRYING

Coal drying is a necessity to allow efficient use of high-moisture coals in combustion or gasification plants. Drying technologies are available for ready implementation at commercial scale. The frontrunners at this stage are the WTA developed by RWE and DryFinding™ developed by Great River Energy. The COMBdry technology is also under rapid development by the Korean Institute of Energy Research and SMK Energy. Development of dewatering technologies such as HTD and MTE appear to have stalled. Also stalled is the development of pressurised steam fluidised bed drying.

Several process simulations and CFD studies have been reported in the literature on retrofitting drying plants to existing coal-fired power stations or predicting the heat flux and temperature distribution inside the boiler. It is clear that firing pre-dried lignite requires modifications to the boiler and the furnace surface cleaning system to deal with the increased fouling. Such modifications will require capital outlay but will be offset by the benefits from reduced coal consumption and reduced flue gas flow.

Modern lignite pre-drying is included in some new technology concepts. For ultra-supercritical lignite-fired power plants using 700°C advanced steam parameters, the plant thermal efficiency gain could reach 7–8% points if pre-drying is incorporated.

Information on the investment cost and techno-economics of drying retrofit is somewhat scarce. Therefore, for any retrofit application, a case-by-case basis analysis of the entire plant is required as the modification requirements are a function of several variables – type, capacity, and design of the boiler, and indeed all the auxiliaries and the moisture content of the coal to be dried. The analysis should not just be based on the process simulation of the full plant, but more importantly also on the heat flux and temperature distribution inside the boiler.

REFERENCES

Atsonios, K., Violidakis, I., Sfetsioris, K., Rakopoulos, D., Grammelis, P., Kakaras, E., (2016). "Pre-dried lignite technology implementation in partial load/low demand cases for flexibility enhancement." *Energy* 96: 165–173.

Burnard, K. and Bhattacharya, S. (2011). *Power Generation from Coal – Ongoing Developments and Outlook*, OECD/IEA, Paris, France.

Cormos, C.-C. (2016). "Oxy-combustion of coal, lignite and biomass: A techno-economic analysis for a large scale Carbon Capture and Storage (CCS) project in Romania." *Fuel* 169: 50–57.

Drosatos, P., Nikolopoulos, N., Nikolopoulos, A., Papapavlou, Ch., Grammelis, P., Kakaras, E. (2017). "Numerical examination of an operationally flexible lignite-fired boiler including its convective section using supporting fuel pre-dried lignite." *Fuel Processing Technology* 166: 237–257.

Fushimi, C. and W. N. Dewi (2015). "Energy efficiency and capital cost estimation of superheated steam drying processes combined with integrated coal gasification combined cycle." *Journal of Chemical Engineering of Japan* 48(10): 872–880.

Global CCS Institute (2018). https://hub.globalccsinstitute.com/publications/post-combustion-carbon-capture-thermodynamic-modelling/33-description-coal-drying-plant, accessed August 10, 2018.

Han, X., Yan, J., Karellas, S., Liu, M., Kakaras, E., Xiao, F. (2017). "Water extraction from high moisture lignite by means of efficient integration of waste heat and water recovery technologies with flue gas pre-drying system." *Applied Thermal Engineering* 110, 442–456.

Kakaras, E., Ahladas, P., Syrmopoulos, S. (2002). "Computer simulation studies for the integration of an external dryer into a Greek lignite-fired power plant." *Fuel* 81(5): 583–593.

Kakaras, E., Koumanakos, A., Doukelis, A., Giannakopoulos, D., Vorrias, I. (2007). "Simulation of a greenfield oxyfuel lignite-fired power plant." *Energy Conversion and Management* 48, 2879–2887.

Kotowicz, J., Balicki, A. (2014). "Enhancing the overall efficiency of a lignite-fired oxyfuel power plant with CFB boiler and membrane-based air separation unit." *Energy Conversion and Management* 80: 20–31.

Lee, S. H. (2018). Korea Institute of Energy Research, personal communication, July 2018.

National Energy Technology Laboratory (2011). "Lignite fuel enhancement – a DOE assessment." DoE/NETL-2011/1490.

Pawlak-Kruczek, H., Lewtak, R., Plutecki, Z., Baranwoski, M., Ostrycharczyk, M., Krochmalny, K., Czerep, M., Zgora, J. (2018). "The impact of predried lignite cofiring with hard coal in an industrial scale pulverized coal fired boiler." *Journal of Energy Resources Technology* 140(6): 062207-1–14.

RWE.com, "WTA Technology – A modern process for treating and drying lignite." https://www.rwe.com/web/cms/mediablob/en/234566/data/213182/6/rwe-power-ag/innovations/coal-innovation-centre/fluidised-bed-drying-with-internal-waste-heat-utilization-wta/Brochure-WTA-Technology-A-modern-process-for-treating-and-drying-lignite.pdf, accessed August 2018.

World Coal (2014). "Drying lignite – how and why?" https://www.worldcoal.com/power/01102014/iea-ccc-report-on-drying-lignite-1381/, accessed August 2018.

Xu, C., Xu, G., Fang, Y., Zhou, L., Yang, Y., Zhang, D. (2014). "A novel lignite pre-drying system incorporating a supplementary steam cycle integrated with a lignite fired super-critical power plant." *Energy Procedia* 61: 1360–1363.

Zero Emissions Platform (2011). "The Costs of CO_2 Capture, Transport and Storage." http://www.zeroemissionsplatform.eu/extranet-library/publication/165-zep-cost-report-summary.html, accessed August 2018.

Zhu, X., Wang, C., Wang, L., Che D. (2018). "Thermodynamic and economic analysis on a two-stage predrying lignite-fueled power plant." *Drying Technology* 1–12.

7 Advances in Coal Drying

Junjie Yan and Xiaoqu Han

CONTENTS

7.1 INTRODUCTION

Fuels with high moisture content, such as low rank coals (LRCs) and some kinds of biomass, are playing an important role in the current global energy market. Total proved reserves for LRC worldwide, including subbituminous and lignite, are 316.7 billion tonnes, accounting for about 30.6% of total coal reserves at the end of 2017 [1].

Lignite is a typical kind of low rank coal. It is formed in the early stages of coal-
ification, with properties falling in between those of bituminous coal and peat. It is
featured by high moisture content (about 20–60%), high ash content (about 6–25%),
low heating value (about 10–21 MJ·kg⁻¹ on an as-received basis), and high volatile
content (about 40–50% on a dry and ash-free basis), as shown in Figure 7.1.

Lignite constitutes a major energy source for many countries and is still in their
electricity planning due to the fact that it has abundant resources and low price, and
provides fuel independence for countries rich in lignite resources. Particularly in
countries such as Germany, Greece, Czech, Poland, Australia, and Indonesia, lignite contributes substantially to the reliability and cost stability of the power supply.
Beside its manifold proclaimed benefits, such as low cost, security of supply, and
grid stabilization impact, lignite has an economic justification. Therefore, in these
countries, lignite is widely used as a source of solid fuel. However, there are environmental concerns caused by the high moisture content of lignite and low efficiencies
in lignite power plants. When a fuel with high moisture content is used to generate
power or heat through a combustion process directly, the evaporation of water will
consume a great amount of high temperature energy from the viewpoint of the second law of thermodynamics [3]. As a result, it is always inefficient in energy utilization when LRCs are used directly.

In order to mitigate the above concerns, LRCs are commonly dried before utilization in some processes. Moreover, coal drying is a necessary process because
of the moisture content limit for various processes. Lignite drying makes it possible to utilize it for power generation more efficiently and environmentally more
friendly [4]. Kakaras et al. [5] in the National Technical University of Athens
conducted simulations for lignite drying systems in 2002, and confirmed that the
integration of drying into lignite-fired power plants resulted in marked efficiency
improvements. From that milestone, a series of studies have been performed for

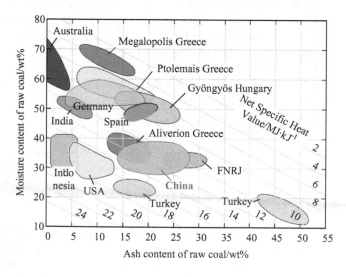

FIGURE 7.1   Comparison of low rank coals worldwide [2].

developing the energy-saving theory of lignite drying and system optimizations. These focus especially on the drying of low rank fuels using low grade heat, and commercial processes have also been demonstrated [6]. Karthikeyan et al. [7] reported the developments of low rank coal drying technologies in 2009. Osman et al. [8] introduced patents and innovations of low rank coal drying technologies until 2011. Jangam et al. [9] provided a critical assessment of industrial coal drying technologies, focusing on the roles of energy, emissions, risk, and sustainability. Zhu et al. [10] and Si et al. [11] paid special attention to steam drying and fluidized drying respectively in 2015. Rao et al. [12] and Nikolopoulos et al. [13] provided comprehensive reviews about developments in drying and dewatering technologies for low rank coals.

The objective of this chapter is to provide an overview of the drying technologies which have been developed in recent years. Discussions on the merits and limitations of drying technologies, their current state, and development direction are presented. This chapter is structured as follows: drying characteristics will be introduced in Section 7.2, in which forms of water in coal, removal process, and key factors of coal quality considered in drying will be covered. Different drying technologies will be provided in Section 7.3, including evaporative and non-evaporative. The origin proposal, working principle, R&D, and industrial applications of these technologies will be reported in detail. Finally, the advances in coal drying will be summarized and reviewed in Section 7.4.

7.2 DRYING CHARACTERISTICS OF LOW RANK COALS

In-depth understanding of the drying process and the materials to be dried is essential for the development and improvement of drying technologies. Therefore, the drying characteristics are introduced firstly in this section.

7.2.1 FORMS OF WATER IN COAL

The ease of water removal is dependent on the forms of water in low rank coals [14]. Water in coal exists in different states, which can be divided into five types: (A) interior adsorption water, (B) surface adsorption water, (C) capillary water, (D) interpartial water, and (E) adhesion water [7], as shown in Table. 7.1 and illustrated in Figure 7.2.

Of the above five types of water, types D and E, also termed *surface moisture*, can be readily removed using *mechanical fine coal dewatering* devices, such as vacuum filters and centrifuges. Type C water can be removed partially as well through mechanical dewatering, depending upon the size of the opening in the coal surface and the drying time available in the filter cycle. *Inherent moisture* is the general term used in a typical proximate analysis of coal to describe types A and B water. This can be removed by *thermal drying processes*, but conventional thermal drying is not the most efficient way of drying all types of water described above. Thus, in order to evaporate water from a large stack of coal, the coal is heated to a high temperature and heat is conducted to the water trapped inside. High temperature results in irreversible changes on coal [7].

138   Drying of Biomass, Biosolids and Coal

TABLE 7.1
Summary of Different Types of Moisture in Coal

Types		Description
A	Interior adsorption water	contained in micropores and microcapillaries within each coal particle, deposited during formation
B	Surface adsorption water	a layer of water molecules adjacent to coal molecules but on the particle surface only
C	Capillary water	contained in capillaries and small crevices found between two or more particles
D	Interparticle water	contained in capillaries and small crevices found between two or more particles
E	Adhesion water	a layer of film around the surface of individual or agglomerated particles

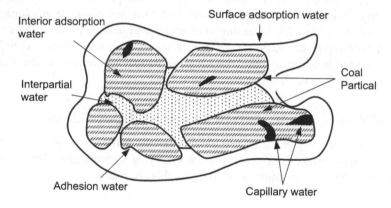

FIGURE 7.2 Type of water associated with coal [7].

The heat of sorption of water on lignite as a function of moisture content for both adsorption and desorption is plotted in Figure 7.3. The isosteric heat of desorption is approximately equal to the heat of vaporization of bulk water at water contents from 200 down to about 40 g per 100 g dry coal. This suggests that the water being desorbed in this region is not bound to the coal; rather, it is water from the spaces between coal particles or in large pores. In other words, vaporization of water from the saturated state to the capillary region of the isotherm is equal to the latent heat of vaporization of pure water. In the region between 40 and 15 g of water per 100 g dry coal the heat of desorption increases gradually. This increase in isosteric heat can be attributed to desorption of water from progressively smaller capillaries or pores. Below water contents of 15 g per 100 g dry coal, the heat of desorption increases sharply. This can be attributed to the desorption of multilayer and then monolayer water which is progressively more strongly bound to the coal surface as the water content decreases [14].

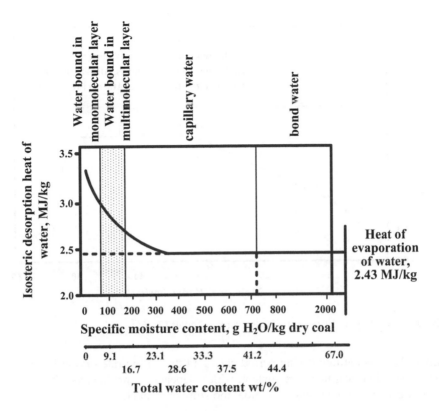

FIGURE 7.3   Heat of sorption of water on lignite as a function of moisture content [15].

7.2.2   DRYING PROCESS

7.2.2.1   Drying Curve

It is also necessary to select the right drying process to meet specific requirements of the intended application. The rate of moisture removal from coal samples and the temperature used will affect the resultant pore structure and therefore properties of the dried material.

Figure 7.4 shows three periods of coal drying. In the initial period, the material is heated up and, after reaching the wet-bulb temperature, drying starts and the drying rate increases until the maximum and constant drying rate is reached. This initial period is followed by a period of constant drying rate, during which the diffusion of water through the solid is sufficiently rapid to maintain saturated conditions at its surface with its evaporation mechanism resembling that of a water body. Finally, the internal diffusion of water can no longer maintain saturated conditions at the particle surface and the drying rate decreases with a decrease in moisture content. The final period may be further divided into two sub-periods for some coals, one sub-period during which the material surface is partially wet but the interior is still wet, and a second sub-period when the material surface is completely dry and the diffusion of water through the solid fully determines the drying rate. The moisture content at

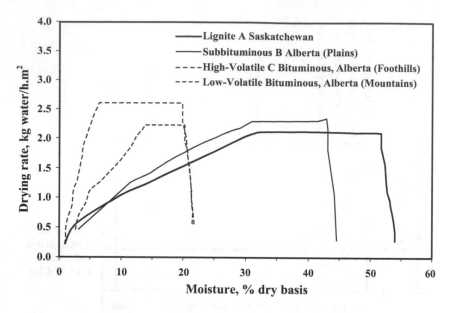

FIGURE 7.4   Drying curves for different ranks of coals [7, 16].

which the constant rate drying period ends and the falling rate drying begins – is the so-called critical moisture content, which depends on parameters like material structure, material thickness, and initial moisture content [13].

7.2.2.2   Mesopore Structure

The variation of pore structure characteristics of low rank coals during water removal is very important and can affect the mass transport and chemical reaction mechanisms involved in combustion and other conversion processes.

The porous structure of low rank coals is mainly comprised of meso- and macropores. The surface area of low rank coals is predominantly controlled by mesopores [17]. Androutsopoulos et al. [18] studied the influences of drying upon macro- and mesopore structures of Greek lignite. During the drying process, considerable particle contraction (by around 1/3 of its original size), a small decrease in macro and partly mesopore volume, and a marked increase in the relevant surface area were observed. They also found the formation of pores in the size range of 7.5–150 nm at the expense of pores in the range 150–1000 nm with the progressive removal of moisture in coal. Pore shrinkage and pore emptying took place as a result of the interaction of particle contraction and moisture removal. Salmas et al. [19] discovered that when drying temperature was raised from ambient condition to 250°C, the mesopore volume and surface area were increased at first and decreased later. The turning temperature was 250°C. Clayton et al. [20] studied the pore destruction caused by mechanical/thermal expression (MTE) dewatering process. It was found that neither an increase in temperature nor an increase in pressure greatly changes the pore diameters that can be collapsed.

7.2.2.3 Moisture Reabsorption

Moisture reabsorption is an important issue in coal production, transportation, and treatment processes, because dried coal is inevitably exposed to ambient humid air.

An increase of around 10–13% in the moisture content of oven-dried coal samples within a period of about two–four days due to reabsorption has been reported by Karthikeyan et al. [21]. The increase varied with the drying temperature and drying methods used. For example, David et al. [22] investigated the drying and reabsorbing characteristics of Victorian lignite in the fluidized bed. It was found that steam-dried coals reabsorbed less moisture than air -dried coals. The average mass difference in moisture was about 1.6%. Li et al. [23] investigated the drying kinetics of an Indonesian low rank coal and moisture reabsorption of the dried coal. It was found that the moisture reabsorption rate was mainly dependent on parameters such as drying temperature, coal particle size and relative humidity of the ambient air. Shen et al. [24] studied the reabsorption of Ximeng lignite from Inner Mongolia, with different drying temperatures, particle sizes, and drying time. It was found that the equilibrium moisture content reached 10–15% in the atmosphere with a relative humidity of 75%. Zhang et al. [25] studied the moisture absorption properties of lignite coals with different moisture content, from Hailaer, Huolinhe and Indonesia, respectively. A critical water content value was obtained to avoid moisture absorption in different conditions.

A serious of upgrading processes have been proposed by far to prevent or minimize the reabsorption of moisture, such as heat treatment at high temperature, coating of dried coal samples using bitumen separately or along with a solvent, and blending of hot dried coal with raw coal samples [26].

7.2.3 FACTORS OF COAL QUALITY AFFECTED BY COAL DRYING

Spontaneous combustion, devolatilization, and combustion behaviors are the main concerns in low rank coal drying, which result from the mesopore structure change.

7.2.3.1 Spontaneous Ignition

Spontaneous ignition of coal occurs naturally without external source of heat as a result of low temperature oxidation. The self-heating is dependent on several factors, such as power plant management, transport management and stockpile storage in silos, bunkers and mills. Moreover, it is also related to controllable features of the coal itself, in particular, the residue moisture content of the dried lignite product [27]. Lignite is especially susceptible to spontaneous ignition due to its high volatile content and porous structure [28]. The drying of lignite seems to aggravate this risk, which is a major concern in storage, handling and transportation of upgraded lignite products. Moreover, the organic matter in low rank coals, also termed young coals, is inherently more reactive than that in older coals, in spite of the generally high moisture content.

However, the effect of moisture content in coal on the propensity for spontaneous ignition is rather complicated. To better understand the process involved, a serious of investigations have been carried out with conflicting results. For instance, Kadioğlu et al. [29] evaluated the spontaneous combustion characteristics of two

142

Drying of Biomass, Biosolids and Coal

types of Turkish lignite after air-drying. The liability of spontaneous combustion was reduced when the residue moisture content increased. On the contrary, Zhang et al. [27] found that the tendency of spontaneous ignition in typical Chinese and Indonesian lignite samples all decreased linearly then the residue moisture content of the dried lignite increased. Zheng et al. [30] reported that lignite coal samples, after high temperature flue gas drying (600–800°C), had a higher ignition activated energy and thereby this hindered self-ignition. Fei et al. [31] compared the spontaneous combustion characteristics of raw and dried Victorian and Pakistan lignite coals. The drying process was completed by mechanical thermal expression, with slurries of the materials heated to 125 or 200°C and pressurized to 5–15 MPa. It was found that drying decreased the critical temperature of the spontaneous combustion of Victorian lignite, but had no effect on Pakistan lignite. After all, it can be stated that the influence of lignite drying on its spontaneous combustion characteristics depends on many factors, including drying methods and temperatures, coal types (volatile and moisture contents), particle sizes, drying time, etc.

7.2.3.2 Devolatilization

Devolatilization of coal is a process in which coal is transformed at an elevated temperature to produce gases, tar and char [32]. Drying of lignite prior to combustion will inevitably induce devolatilization. Drying temperature is the key parameter that determines devolatilization. In superheated steam drying, it was found that the release of organic matters became significant only when the greater part of water had been evaporated [33]. Organic matters released during superheated steam drying have been shown to consist predominantly of a mixture of phenol and methoxyphenol. In flue gas drying, a very small amount of volatiles would be generated even in relatively harsh dry conditions, as reported by Lv et al. [34] The increases of particle size and initial moisture content can prevent lignite devolatilization during high temperature flue gas drying [35]. In vacuum drying [17], devolatilization took place at 200°C and was significantly increased with further increases in the drying temperature.

7.2.3.3 Combustion Behavior

The residue moisture content of lignite after drying is close to that of bituminous coal, and its combustion characteristics are important for the design of lignite boilers. Agraniotis et al. [37] conducted experimental and theoretical investigations on the combustion behavior when raw and dried lignite were co-fired. Ahmed et al. [38] conducted numerical investigations on fluid flows and combustion behaviors of raw lignite and mechanical/thermal expression (MTE) upgraded lignite firing scenarios. Zhao et al. [39] investigated the combustion of dried lignite in a 375 MW tangentially fired furnace, and recommended the use of recirculated flue gas as an option for future operation of the furnace with dried coal. Man et al. [40] investigated the combustion and pollutant emission characteristics of lignite dried by low temperature air (120–180°C), and found that the burnout temperatures of the dried coals were higher than those of the raw coal, whereas ignition temperatures kept nearly constant. Wang et al. [41] investigated the impact of staged air on NO_x emissions in the combustion of dried lignite by experiments. It was found that NO_x emissions decrease with increasing blending ratios of dried coal. In summary, further research work is required on combustion characteristics of dried lignite towards fouling and slagging mitigation, pollutant and emissions reduction, and operation optimization in full working conditions.

7.3 DRYING TECHNOLOGIES FOR LOW RANK COALS

7.3.1 CLASSIFICATION OF DRYERS

A plenty of water removal technologies for lignite have been proposed until now. Therefore, different categories of dryers have been defined according to certain parameters: the drying method, controlling heat transfer mechanism, heating and drying media, heat source, and pressure. In general, these technologies can be classified into two types from the perspectives of thermal processes: *evaporative drying* and *non-evaporative dewatering*. In the evaporative drying process, the moisture is transformed into the gaseous phase (as steam), while in the non-evaporative dewatering processes the moisture is removed as a liquid. The energy consumption for the latter is generally lower than the former since the latent heat of vaporization is avoided.

7.3.2 EVAPORATIVE DRYING TECHNOLOGIES

Evaporative dryers include fixed beds [42], fluidized beds, rotary kilns, moving beds, and other emerging technologies. Representative evaporative drying technologies are introduced in this section.

7.3.2.1 Rotary Dryer

Rotary drying technology is relatively mature and has been widely applied in many fields. According to the form of contact between coal and drying medium, rotary dryers can be of either direct or indirect type. Rotary-drum and rotary-tube drying systems are the most common rotary dryers, as shown in Figure 7.5.

7.3.2.1.1 Rotary-Drum Dryer

The rotary-drum drying system can be both direct and indirect contact. In the former type, high temperature hot flue gas (generally ≥350°C), is in contact with coal in the drum and the water is removed by convective heat transfer. Although there exist fire risks, this type of drying is commonly used in industrial occasions. In the indirect-contact rotary-drum dryer, tubes (one–three laps) are arranged concentrically in the drum. The hot flue gas at high temperature flows through the tubes and heats the drum chamber and the coal in contact with the outer surface of the tubes. Since the coals do not come in direct contact with the heat transfer medium, steam can also be used instead of flue gas. Generally, the rotary-drum dryer presents an obvious constraint on the particle size, usually lower than 30 mm [43]. Lignite with an initial moisture content of 34% can be dried to less than 10% in the rotary drum dryer [44]. Moreover, the heat transfer coefficient obtained in a lab scale experiment varied from 33 to 37 W·m^{-2}·K^{-1} [45].

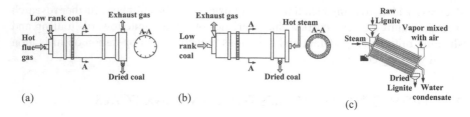

FIGURE 7.5 Schematic diagrams for different types of rotary drying systems [43]. (a) Direct rotary drum dryer (b) Indirect rotary drum dryer (c) Rotary-tube dryer

7.3.2.1.2 Rotary-Tube Dryer

The rotary-tube drying system (Figure 7.5c), developed by ZEMAG Clean Energy Technology GmbH, is an improved indirect-contact rotary-drum dryer. Coals are fed to the inclined tubes and heated by the drum chamber as well as the tube surfaces while the drying medium flows through the drum chamber outside the inclined tubes. With the rotary motion, raw low rank coals are continually transported to the exit. The thermal energy required for the moisture evaporation is supplied by low pressure steam which enters the dryer drum along the axis. The temperature of hot steam used in the rotary-tube drying system is usually lower than 200°C, which reduces the risk of self-ignition. Hatzilyberis et al. [46] predicted the overall heat transfer coefficients of a pilot rotary-tube dryer as 75 $W \cdot m^{-2} \cdot K^{-1}$ (by experiment), and 78 $W \cdot m^{-2} \cdot K^{-1}$ (by modeling). According to the investigation results of Feng et al. [47], the heat transfer coefficient of the rotary-tube dryer could reach 94 $W \cdot m^{-2} \cdot K^{-1}$ when Indonesian lignite was dried.

The main limitations of rotary dryers are high energy consumption (3700 kJ/kg H_2O) and high capital and maintenance costs.

7.3.2.2 Mill Type Dryer

The mill type dryer combines the grinding and drying operations of coals, which is a standard practice in pulverized lignite-fired power plants. In this case, the drying medium comes from the top of the furnace, i.e. the hot flue gas (~1000°C) is being recirculated to the milling system which consists of a number of beater mills. The mixture of hot flue gas, primary air and evaporated moisture also functions as a carrier gas to feed the pulverized lignite particles to burners. In standard systems, all evaporated moisture is recirculated to the boiler along with the lignite particles. Nevertheless, for lignite coals with very high moisture and ash contents, it is necessary to separate part of the moisture. This type of system has been applied in the Megalopoli power plant in Greece and the Elbistan power plant in Turkey for efficiency improvement and flame stability (Figure 7.6).

7.3.2.3 Moving Bed Dryer

The packed moving bed dryer was developed to address the need for large heat capacities in lignite drying. Advantages of the dryer include the use of a lower temperature flue gas source and the waste heat recovery. For example, flue gas at a low temperature (150°C) and oxygen volume content (less than 8%) was used in woody

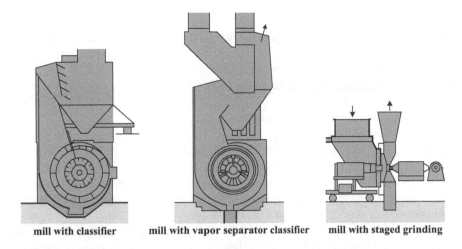

mill with classifier   mill with vapor separator classifier   mill with staged grinding

FIGURE 7.6 A schematic of the mill type dryer [4].

biomass drying for bioenergy [48]. Usage of the exhaust gases at this temperature levels makes it possible to recover waste heat. Besides, the spontaneous combustion risk can be decreased because of the controlled oxygen content. In this regard, the methodology seems applicable to power plants with high capacities and waste heat recovery systems. Numerical simulations and experimental studies of lignite drying in a packed moving bed dryer were conducted by Zhang et al. [49], for insights into key parameters that influence the drying efficiency. The parameters included drying gas velocity, lignite particle size, lignite particle moving velocity, and lignite particle stack height.

7.3.2.4 Fluidized Bed Dryer

The concept of steam fluidized bed drying (SFBD) was proposed by Potter at Monash University in Australia [50]. A high drying rate can be achieved in FBDs due to an adequate mixing, use of a high temperature heating (and fluidizing) medium, and ease of control. A fluidized state is achieved when drying medium flows through the coal layer at a proper velocity. The gas flow velocity at which the packed bed is converted into a fluidized bed is known as the minimum fluidization velocity. Jeon et al. [51] experimentally studied the drying characteristics of lignite in bubbling fluidized bed with coal particle sizes ranging from 0.3 to 2.8 mm using. It was found that the drying process can be completed within 10 min. Wang et al. [52] presented a laboratory scale fluidized bed drying of three different types of Illinois coals with initial moisture contents of 18, 20, and 23%, respectively. The effects of some parameters, such as drying temperature, on the drying performance were investigated. Kim et al. [53] analyzed the drying behavior of Australia's Loy Yang lignite in a fluidized bed dryer. It was shown that the drying rate increased with increasing drying temperature and higher fluidization velocity, and decreasing relative humidity.

Until now, several types of fluidized bed dryers have been developed and some of them are under demonstration.

7.3.2.4.1 WTA Dryer

RWE Power in Germany developed the WTA technology (German abbreviation: Wirbelschicht-Trocknung mit interner Abwärmenutzung), which uses internal low temperature heat for coal drying in the fluidized bed dryer. It is arguably one of the most advanced superheated stream drying techniques. It was reported that the WTA process consumes 80% less energy compared to rotary steam drying with 80% less dust emission and lower capital investment.

In the applied WTA process, lignite should be primarily milled to a fine particle size by hammer mills in direct series with a two-stage fluidized bed dryer. The dried fuel exiting the stationary bed is separated from the gas stream, It is then mixed with coarser lignite solids collected from the bottom of the dryer bed and fed to the furnace. To recover the latent heat, two integration concepts were implemented, including vapor condensing (Figure 7.7a) and compression (Figure 7.7b).

In the former type, the heat required for lignite drying is supplied by external steam extracted from the turbine. In the latter type, vapor compression is applied to recover the latent heat from the humid dryer exhaust gas. The heat exchange is realized in tube bundles located inside the bed.

The WTA process (Figure 7.7a) has been demonstrated in the Niederaussem plant in 2009, where 25% of Unit K's input fuel is being treated. The coal moisture content is decreased from 50% to between 10 and 18% prior to feeding into the mills, as designed. Consequently, the overall efficiency of the plant is raised by 1 percentage-point. The vapor compression system (Figure 7.7b) has also been applied successfully in the Frechen plant.

7.3.2.4.2 Pressurized Fluidized Bed Dryer

The pressurized fluidized bed dryer (PFBD), developed by Vattenfall, is similar in principle to WTA, except that it operates in a higher fluidized bed pressure. According to the extended work of Hoehne et al. [55] on the PSFBD, the mean heat transfer coefficient with internal heating could be 250–300 $W \cdot m^{-2} K^{-1}$. The influence of steam pressure, velocity, coal type and particle size on the heat transfer coefficient was studied. It was also found that high heat transfer coefficient is advantageous for the removal of water at low moisture content because a high temperature is required to remove the water in the capillaries [56]. Lechner et al. [57] have completed the high pressure superheated steam drying of lignite on a pilot scale (Figure 7.8).

7.3.2.4.3 Low-Temperature Fluidized Bed Dryer

A low-temperature fluidized bed drying method, the DryFining™ system (Figure 7.9), which uses hot air as the drying/fluidized medium and cooling water from the condenser as the heating medium, was proposed by Great River Energy. The in-bed heat exchangers are used to increase the temperature of the air and its moisture-carrying capacity. Therefore, it can also use low grade waste heat available in the plant. Moreover, it decreases the risk of oxidation and fire occurrence owing to the usage of low-temperature air as a drying medium. Sarunac et al. [58] reported that when the moisture content of the lignite reduced from 38.5 to 23.5% through the integration of DryFining™, the plant net efficiency could be improved by 1.7%.

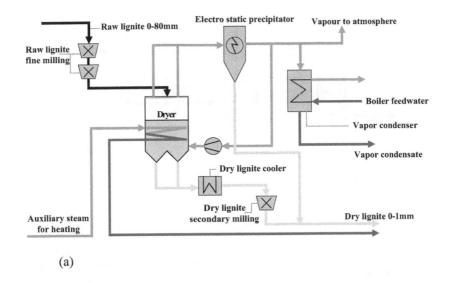

(a)

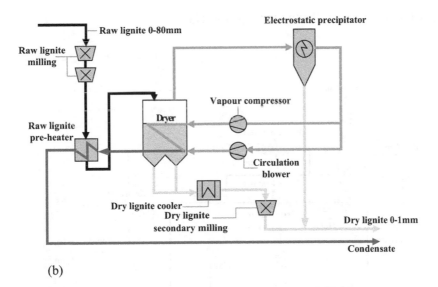

(b)

FIGURE 7.7 Schematic diagrams of the WTA drying concept. (a) WTA concept with vapor condensation in RWE Power Niederaussem power station and (b) Process principle of WTA fine-grain drying with vapor recompression [54].

7.3.2.4.4   Self-Heat Recuperation-Based Continuous Fluidized Dryer

To reduce energy consumption for drying and dewatering, a self-heat recuperation-based continuous fluidized bed dryer (SHR-FBD) for drying lignite has been developed by Aziz et al. [59]. It is claimed that this dryer is capable of recovering both latent and sensible heat from the evaporated water, and sensible heat from dried products effectively. The schematic diagram of the SHR-FBD is shown in Figure 7.10.

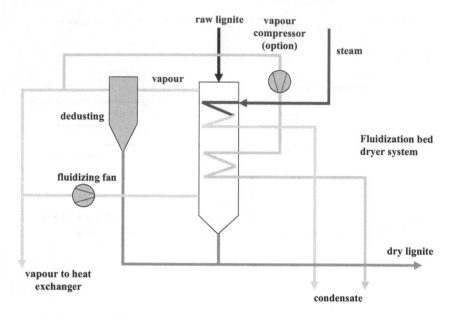

FIGURE 7.8   A schematic of the PFBD process [13].

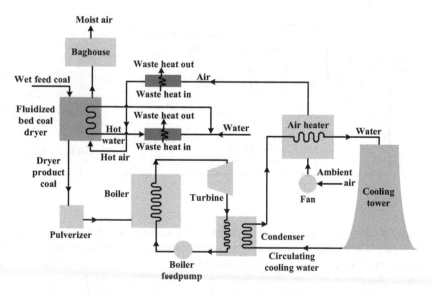

FIGURE 7.9   Drying Fining™ system by the GRE [58].

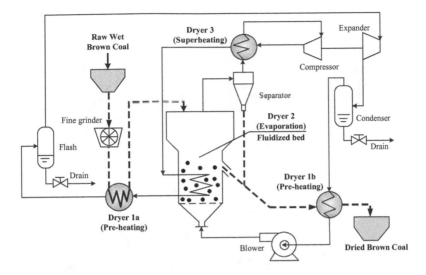

FIGURE 7.10   A schematic of the SHR-FBD system [59].

The drying process includes three continuous stages: preheating, evaporation, and superheating. Latent heat can be recovered from the exhaust drying medium and utilized in the drying system, which can improve the overall energy efficiency. In contrast to the drying system integrated with conventional heat recovery methods, the energy consumption in the SHR-FBD system showed a significant reduction by approximately 70% [60].

7.3.2.4.5   Spouted Bed Steam Dryer

The spouted bed dryer is a special modification of the FBDs, in which a high-velocity jet of gas penetrates through an opening at the bottom of the bed of particles and transports the particles to the bed surface. In this way, an intensive solid mixing is induced. A pilot-scale superheated steam spouted bed fluidized dryer was designed and constructed in Karlstad University, Sweden [61]. A diagram of the demonstration-scale drying plant is shown in Figure 7.11. The dryer works under atmospheric pressure, and the superheated steam temperature was 201°C. The wet material, including sawdust and willow wood chips, was fed in and the moisture content was reduced from 51.9 to 19.4%.

The spouted bed has been applied to granulation, coating and drying of pastes, solutions, slurries, and suspensions.

7.3.2.4.6   Vibrated Fluidized Bed Dryer

Mechanical vibration of the fluidizing medium can help improve the performance of fluidized bed drying. It has been reported that vibrated bed dryers have better energy efficiency compared to conventional fluidized bed dryers [6, 7]. The minimum fluidization velocity is relatively lower than that of the traditional fluidized bed dryer [62]. Higher amplitude and frequency of vibration would lead to greater material flow

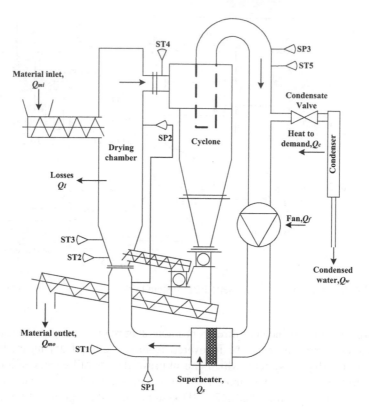

FIGURE 7.11   Diagram of the spouted bed steam dryer [61].

ability [63]. Moreover, attrition and gas cleaning requirements are also minimized using a vibrated dryer. Take a commercial vibratory dryer for hard coal and lignite manufactured by Escher-Wyss of Switzerland for example. It used a vibration frequency of 50–100 Hz and amplitude of 0.5–3 mm, reaching a conveying velocity in the range of 0.01–0.3 m/s with an angle of inclination of 5° to the horizontal. Low gas velocities are needed because vibration suspends most of the pseudo-fluidized beds.

7.3.2.5   Microwave Dryer

Microwave energy is non-ionizing electromagnetic radiation with frequencies from 300 MHz to 300 GHz, or wavelengths from 1 to 300 mm [12]. Due to the high absorptive capacity of water to microwave energy, microwave drying has been employed in a few coal drying studies. It has significant advantages such as volumetric, rapid and selective heating, uniform heat distribution, faster drying rates compared to the conventional heating, possibility of using intermittent exposure of wet solids to microwaves, flexible modular design, environmentally friendly application, fast switch-on and switch-off, high efficiency, etc. Hence, the original purpose of microwave heating is used to enhance the magnetic removal of pyrite within coals. With high heating rates and reduced processing time, microwaves have been widely used for drying of low rank coals, such as the CoalTek process developed by CoalTek Inc., and the Drycol process (Figure 7.12a) developed by DBAGlobal Australia Pty Ltd.

During microwave drying, three periods can be observed: preheating period, fast drying rate period, and falling rate drying period (Figure 7.12b). Microwave radiation is found to be more applicable for drying coarse lignite with large particle sizes [66–68]. But it seems very costly for low rank coal drying in comparison with other drying technologies, since electricity is directly used as energy source. An alternative method is to integrate the microwave drying with a conventional drying system so that the advantage of microwave drying in removing water inside the coal, which is difficult to evaporate with the other drying technologies, can be taken. As a result, drying time can be shortened evidently.

The main limitations of microwave dryers include hot spots due to the presence of impurities, fire hazards resulting from high dielectric losses, and high cost involved in handling large amounts.

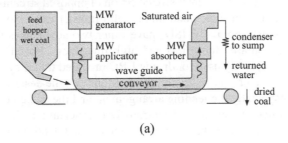

(a)

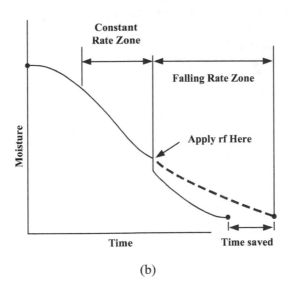

(b)

FIGURE 7.12   (a) Schematic of the Drycol process [13, 64] and (b) Comparison of typical microwave and dielectric drying curves [65, 66].

7.3.2.6 Pneumatic Dryer

Pneumatic drying, also known as flash drying, is an application of circulating fluidization and dilute transport in drying. Direct contact of the fine coal particles with the heating medium allows rapid drying. Superheated steam, hot air, or exhaust flue gases can be used as the drying medium, with the temperature of the dryer inlet as high as 550–700°C and the corresponding outlet temperature of 70–170°C. The technology requires high drying medium velocities to transport the coal particles, with sizes ranging from 0.01 to 0.5 mm. It is therefore not applicable for large particle sizes. The application of pneumatic drying is the BCB (binderless coal briquetting) developed by White Energy Co. ltd in Australia, and the HPU (hot press upgrade) developed by Shenhua Co. ltd in China (Figure 7.13).

7.3.2.7 Impinging Stream Dryer

Impinging stream dryers (ISDs) are novel alternatives to flash drying for particulate materials with very high drying loads. A schematic diagram for this drying system is illustrated in Figure 7.14. An intensive collision of opposed streams creates a zone that offers very high heat, mass, and momentum transfer rates. So it is possible to remove surface moisture rapidly. ISDs have other two advantages due to no moving parts required: smaller footprints and high robustness. Based on these advantages, ISDs can be possibly used for drying LRC provided that they can handle high throughputs, which can be the main limitation. Moreover, the design of such a system is very critical; especially the feeding arrangement and the design of the impinging pipes affect the value of volumetric heat transfer coefficient and, in turn, the drying rate. Another limitation could be the high velocities used. Until now, studies on ISD are still partial or limited to very few applications. Choicharoen et al. [69] carried out performance evaluation of an impinging dryer with okara as an ideal material and concluded that ISD gives a very high volumetric heat transfer coefficient.

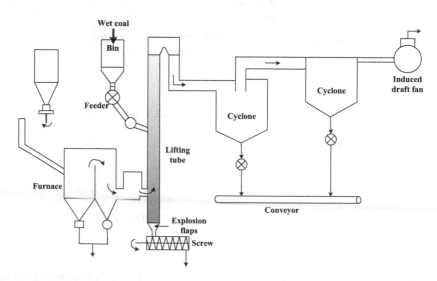

FIGURE 7.13 A schematic of the pneumatic dryer [13].

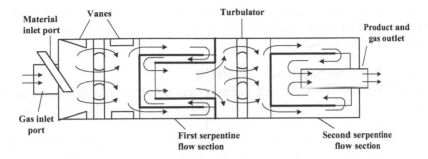

FIGURE 7.14 A schematic of the ISD dryer [8].

7.3.2.8 Screw Conveyor Dryer

The screw conveyor dryer (SCD) comprises a jacketed conveyor in which material is simultaneously heated and dried as it is mechanically conveyed (Figure 7.15) [70]. The heating medium is usually hot water, steam, or a high temperature heat transfer medium such as pot oil, fused salt, or Dowtherm heat transfer fluid. The advantages of SCDs include the applicability for indirect heating, small dryer size, and high thermal efficiency. Setting up the SCD in a multistage drying system can also be an option for higher drying performances. The main limitations of SCDs are high maintenance cost and abrasion caused by moving parts.

7.3.3 NON-EVAPORATIVE DEWATERING TECHNOLOGIES

Compared with evaporative drying, non-evaporative dewatering is more energy efficient. Through non-evaporative dewatering techniques, such as hydrothermal

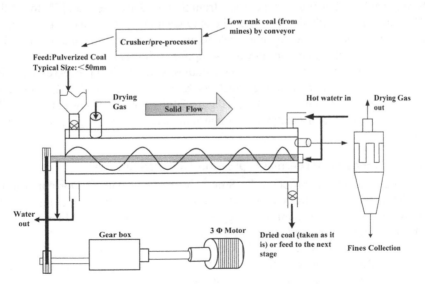

FIGURE 7.15 A schematic of the SCD dryer [9].

dewatering and mechanical/thermal expression dewatering, the pore structure of low rank coals can be changed and the calorific value of low rank coals can be upgraded to that of bituminous coals.

7.3.3.1 Hydrothermal Dewatering (HTD)

In the HTD process, coal is heated under pressure (~3 MPa) to reach temperatures in the range of 250–350°C. Under these conditions, the coal structure breaks down and shrinks, and the water is released as a liquid.

A hydrothermal batch-type reaction system was designed by Mursito et al. [71] for dewatering raw tropical peat. The optimum temperature for hydrothermal treatment was found to be 250°C. Using the HTD method, Favas et al. [72] concluded that the intra-particle porosity of the dried products was mainly affected by the reaction temperature, which was optimized as around 320°C for dewatering Loy Yang low ash Victorian lignite. In a further study, Favas et al. [73] compared the low rank coals from Australia, Indonesia, and the U.S., and found that rank was the major factor influencing the properties of HTD products. Besides, Racovalis et al. [74] investigated the effect of processing conditions on organics in wastewater from HTD treatment of Loy Yang and Yallourn lignite. It was found that the concentration of organics increased exponentially when the temperature increased from 250 to 350°C. Sakaguchi et al. [75] proposed the usage of the water inherent in lignite instead of adding extra water to realize HTD treatment. It was found that the water content was reduced from 59 to 6% and the absolute heating value of the upgraded coal was comparable to or even higher than that of the raw coal during treatment at 350°C, which makes HTD more energy efficient and water saving. Yu et al. [76] prepared coal water slurry from two types of lignite, in XiMeng and BaoTou in China. The lignite coals were dewatered under HTD conditions at 320°C. It was shown that the maximum solid concentration of coal water slurry prepared from XiMeng coal increased from 45.7 to 59.3%, whereas that from BaoTou coal increased from 53.7 to 62.1%. Ge et al. [77] found that HTD could significantly decrease the inherent moisture and oxygen content, increase the calorific value and fixed carbon content, and promote the damage of the hydrophilic oxygen functional groups in Chinese lignite with various ranks.

Based on theoretical and experimental studies, several HTD processes are being developed for commercial market, such as K-fuel [4] (Figure 7.16a), continuous hydrothermal dewatering (CHTD) [4] (Figure 7.16b), hot water dewatering (HWD) [78] (Figure 7.16c), and catalytic hydrothermal reactor technology (Cat-HTR) [4] (Figure 7.16d).

7.3.3.2 Mechanical/Thermal Expression Dewatering (MTE)

The MTE concept was proposed by Strauss at the University of Dortmund in the mid-1990s to improve the energy efficiency of power plants firing lignite [79]. Thermal energy and mechanical energy are combined during the MTE process to enable rapid dewatering and reduce investment cost.

MTE squeezes the water out of lignite with mechanical energy, typically at pressures of 13–25 MPa and temperatures of 150–200°C. In these systems, the dewatering process is usually carried out in four steps (Figure 7.17): 1) fitting the lignite into the chamber with additional water, 2) heating the coal to a set temperature for an

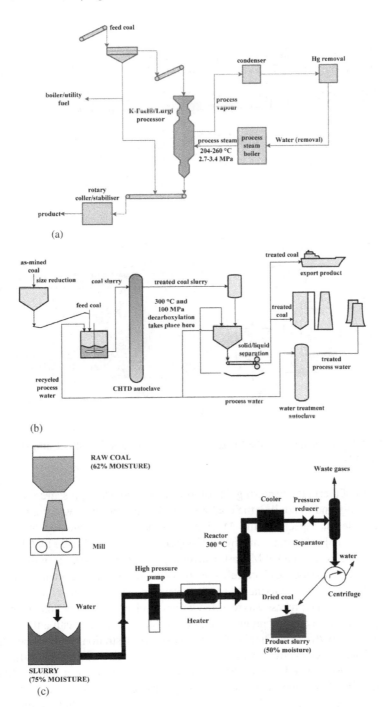

FIGURE 7.16 Different concepts of HTD in commercial development. (a) Flowsheet of the K-fuel® Process [4], (b) Flowsheet of the CHTD Process [4], (c) Flowsheet of the HWD concept [13,73], and (d) Cat-HTR process diagram [4].

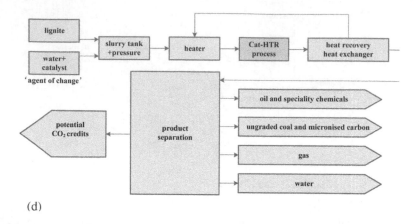

(d)

FIGURE 7.16   (Continued)

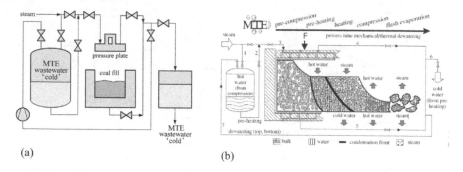

(a)   (b)

FIGURE 7.17   A schematic of the MTE concept. (a) Process flowsheet and (b) Process design [80, 81].

equilibrium of thermal dewatering, 3) compressing to constant pressure for mechanical dewatering, and 4) cooling the coal with flash evaporation dewatering.

Many different aspects of the MTE process has been researched at Dortmund University [82, 83] in Germany and by the Cooperative Research Centre for Clean Power from Lignite (CRC Lignite) at Monash University in Australia [84]. From these studies, it has been illustrated that MTE can effectively reduce the moisture content of Australian and other lignite coals. In Germany, a further development towards commercial implementation of the MTE process was reported which involved the conversion of the discontinuous pilot press to quasi-continuous fully automatic operations, with a throughput of approximately 1.6 t/h of dried lignite. A 25 t/h MTE demonstration plant was constructed at RWE's Niederaussem power plant and was commissioned at the end of 2001, but was subsequently discontinued. In the CRC process, the coal was fed as a slurry, which was preheated using energy recovered from the hot product coal and hot expressed water.

Various aspects of the MTE technology have been tested, including: (a) the feeding of the MTE press with coal under quasi-continuous conditions, (b) the treatment of the raw lignite, and (c) the subsequent dry lignite treatment. The liquid water removed from coal by HTD and MTE processes carries with it both organic and inorganic matter. The large volumes of acidic, salty, and organic-rich wastewater present a major concern in terms of wastewater treatment difficulties and costs for disposal or reuse. The overall viability of the processes will depend on the availability of a simple and energy efficient water remediation strategy. Investigations into ways of treating the wastewater from HTD and MTE processes were increasing and more work in this aspect is required. For example, water remediation using the feed coal itself as an adsorbent may provide such an option [85].

7.3.3.3 Solvent Extraction

The conceptual method of solvent extraction was developed by Miura et al. [86]. More specifically, water is released into the solvent through heating; the mixture of water and solvent can be removed from the coal, and then the mixture is separated by cooling to the ambient temperature. The dewatering process by solvent extraction can be carried out at very low temperatures, making it energy efficient.

Miura et al. [86] designed an apparatus for experimental testing of solvent extraction. When tetralin was used as the solvent with treating temperature of 150°C, the water content of Australian lignite samples, weighing 200–300 mg, can be reduced from 50 to 2%. Fujitsuka et al. [87] tried to use non-polar solvents, such as 1-methylnaphtalene, to dewater and upgrade low rank coals. The inherent water was confirmed to be almost completely removed from the coal without phase change and could be separated from the solvent by decantation at room temperature. Moreover, solvent treatment is also an effective method for suppressing low temperature oxidation reactivity of low rank coals. However, this technology was tested only in lab scale.

Liquefied DME can also be used as a solvent. Kanda et al. [88] performed laboratory-scale experiments for dewatering subbituminous coal mined in Warra, Indonesia using DME. It was found that the maximum water extraction efficiency of liquefied DME was 98.3% and the properties of the coal showed no change after the process. Therefore, the liquefied DME is also confirmed effective and energy efficient for coal dewatering.

S-CO_2 has been widely used for coal dewatering, which is most promising at near-room temperatures. After being dried with S-CO_2, coals show no shrinkage or collapse of pore structure. Iwai et al. [89] used S-CO_2 for dewatering Berau lignite and found that the residue moisture content could be reduced to lower levels by adding methanol to S-CO_2.

The challenge of the solvent extraction method is the uncertainty of performance at a larger scale as no report has been found in literature on scaled-up coal dewatering using solvent extraction technology.

7.3.4 COMPARISON OF TYPICAL DRYERS

Various drying methods have been discussed above. Generally, there is no universal dryer type suitable for every case, but the choice of the dryer depends on the particular conditions and the intended application. The main concerns for LRC upgrade and optimization are to minimize the overall energy consumption in LRC drying, while making the drying process cost effective. Although available data concerning all types of industrial dryers is still lacking, a comparison of typical evaporative dryers, as shown in Table 7.2, has been presented by Nikolopoulos et al. [13].

TABLE 7.2

A Summary of Typical Range of Design and Performance Specifications, Capital Cost, and Operating Consumptions for Various Types of Dryers [13]

	Rotary Dryer	Pneumatic Dryer	Fluidized Bed Dryer
Specific consumption (MJ/kg-evaporation)	3.0–4.0	2.7–2.8	2.2[a]–3.5
Drying efficiency (%)	50–75	75–95[a]	70–90[a]
Evaporation (t/h)	3–23	4.8–17	2–41
Drying medium temperature (°C)	200–600	150–280	160–280
Capacity (t/h)	3–45	4.4–16	1.5–25
Feed moisture (%)	45–65	45–65	45–65
Moisture discharge (%)	10–45	10–45	10–45
Typical residence time within dryer	5–10 min	0–10 s	5–10 min
CAPEX (€/dry kg/h)	120–290	260–680	160–285
Operating electrical consumption (kWh/dry t)	8–14	16–38	15–20

[a] With heat recovery.

7.4   SUMMARY AND OUTLOOK

Drying of low rank coals prior to utilization, with appropriate drying technologies, has proven an effective method to upgrade the fuel, improve plant efficiency, and reduce CO_2 emissions. Various technologies for low rank coal drying have been developed or are under development. They can be classified into two categories: evaporative drying and non-evaporative dewatering processes. This chapter focused on the advances in process design and industrial applications.

Typical evaporative drying methods include rotary dryer, mill type dryer, fluidized bed dryer, microwave dryer, pneumatic dryer, impinging stream dryer and screw conveyor dryer. Among them, rotary drying and fluidized bed drying are the two most common methods and have been applied successfully in the industry. Rotary drying, with hot air, flue gas, or steam as a drying medium, is currently technically mature; however, its disadvantages include high energy consumption, high capital costs, and limited processing capacity. Fluidized bed drying, using superheated steam or hot air as the drying and fluidizing medium, is widely applicable in lignite drying for its high drying rate, high processing capacity, and low maintenance cost.

There are three main types of non-evaporative dewatering processes: hydrothermal dewatering (HTD), mechanical thermal expression (MTE), and solvent extraction. MTE, a typical non-evaporative drying method that combines the mechanical expression and the thermal dewatering processes, is reported to have low energy consumption but is not available at industrial scale. HTD is also an energy efficient non-evaporative drying method but is not available for large scale application either. Besides, new concepts of drying methods are emerging continuously. These novel technologies provide diverse options for users, but they must be further researched and improved before large-scale applications. Moreover, the wide use of other drying and dewatering technologies will be significantly determined by the economics of the global coal market.

In order to minimize energy consumption during drying of low rank coals, three aspects should be considered:

1. Low grade heat source utilization and energy matching in process design. Efforts should be made on using energy sources with low exergy as a drying heat source in order to reduce utilization of plant hot utility. Based on the fundamentals of thermodynamics, exergy reflects the quality of energy that is equivalent to work. Therefore, the second-law efficiency of a drying system could be improved when energy low in exergy is used as drying heat source. Typically, the recovery of waste heat from boiler exhaust gases and turbine exhaust steams of the power units is essential, especially when the drying system is inside the power plant.
2. Waste heat utilization. Exhaust gas from dryer possesses a certain quantity of energy depending on the exhaust gas temperature. Efforts should be put on the recovery and reuse of the waste heat in the dryer exhaust gas for other processes or for heat pump drying.
3. Drying process intensification. For the drying and dewatering systems close to the source of coal (such as a mine shaft), it is necessary to improve the thermal efficiency of the drying process itself. Heat transfer enhancement in dryers needs to be carried out.

ACKNOWLEDGMENTS

The authors' work was supported by the National Basic Research Program of China (973 Program, Grant Number 2015CB251504), and the National Natural Science Foundation of China (Grant Number 51436006, 51806159).

REFERENCES

1. Global BP. BP statistical review of world energy. 2018.
2. http://earthresources.vic.gov.au/earth-resources/victorias-earth-resources/coal [accessed in July 2018].
3. Han XQ, Liu M, Wu KL, et al. "Exergy analysis of the flue gas pre-dried lignite-fired power system based on the boiler with open pulverizing system". *Energy*, 2016, 106: 285–300.
4. Zhu Q. "Update on lignite firing". 2012. Available on: https://www.usea.org/sites/default/files/062012_Update%20on%20lignite%20firing_ccc201.pdf [accessed in July 2018].
5. Kakaras E, Ahladas P, Syrmopoulos S. "Computer simulation studies for the integration of an external dryer into a Greek lignite-fired power plant". *Fuel*, 2002, 81(5): 583–593.
6. Mujumdar AS. *Handbook of Industrial Drying*. CRC Press, Boca Raton, FL 2014.
7. Karthikeyan M, Wu ZH, Mujumdar AS. "Low-rank coal drying technologies – Current status and new developments". *Drying Technology*, 2009, 27(3): 403–415.
8. Osman H, Jangam SV, Lease JD, et al. "Drying of low-rank coal (LRC) – A review of recent patents and innovations". *Drying Technology*, 2011, 29(15): 1763–1783.

9. Jangam SV, Karthikeyan M, Mujumdar AS. "A critical assessment of industrial coal drying technologies: Role of energy, emissions, risk and sustainability". *Drying Technology*, 2011, 29(4): 395–407.

10. Zhu J, Wang Q, Lu X. "Status and developments of drying low rank coal with super-heated steam in China". *Drying Technology*, 2015, 33(9): 1086–1100.

11. Si CD, Wu JJ, Wang Y, et al. "Drying of low-rank coals: A review of fluidized bed technologies". *Drying Technology*, 2015, 33(3): 277–287.

12. Rao ZH, Zhao YM, Huang CL, et al. "Recent developments in drying and dewatering for low rank coals". *Progress in Energy and Combustion Science*, 2015, 46: 1–11.

13. Nikolopoulos N, Violidakis I, Karampinis E, et al. "Report on comparison among current industrial scale lignite drying technologies (A critical review of current technologies)". *Fuel*, 2015, 155: 86–114.

14. Yu JL, Tahmasebi A, Han YN, et al. "A review on water in low rank coals: The existence, interaction with coal structure and effects on coal utilization". *Fuel Processing Technology*, 2013, 106(2): 9–20.

15. Allardice DJ. "The water in brown coal" [D]. PhD thesis, University of Melbourne, Australia, 1968.

16. Rozgonyi TG, Szigeti LZ. "Upgrading of high moisture content lignite using saturated steam". *Fuel Processing Technology*, 1985, 10(1): 1–18.

17. Mahidin, Ogaki Y, Usui H, et al. "The advantages of vacuum-treatment in the thermal upgrading of low-rank coals on the improvement of dewatering and devolatilization". *Fuel Processing Technology*, 2003, 84(1–3): 147–160.

18. Androutsopoulos GP, Linardos TJ. "Effects of drying upon lignite macro-pore structure". *Powder Technology*, 1986, 47(1): 9–15.

19. Salmas CE, Tsetsekou AH, Hatzilyberis KS, et al. "Evolution lignite mesopore structure during drying – Effect of temperature and heating time". *Drying Technology*, 2001, 19(1): 35–64.

20. Clayton SA, Wheeler RA, Hoadley AFA. "Pore destruction resulting from mechanical thermal expression". *Drying Technology*, 2007, 25(4): 533–546.

21. Karthikeyan M, Kuma JVM, Chew SH, et al. "Factors affecting quality of dewatered low rank coals". *Drying Technology*, 2007, 25(10): 1601–1611.

22. Stokie D, Woo MW, Bhattacharya S. "Comparison of superheated steam and air fluidized-bed drying characteristics of Victorian brown coals". *Energy & Fuels*, 2013, 27(11): 6598–6606.

23. Li XC, Song H, Wang Q, et al. "Experimental study on drying and moisture re-adsorption kinetics of an Indonesian low rank coal". *Journal of Environmental Sciences*, 2009, 21(S1): 127–130.

24. Shen WJ, Liu JZ, Yu YJ, et al. "Experimental study on drying and reabsorption of the lignite of Ximeng". *Proceedings of CSEE*, 2013, 33(17): 64–70 [in Chinese].

25. You CF, Wang HM, Zhang K. "Moisture adsorption properties of dried lignite". *Energy & Fuels*, 2012, 27(1): 177–182.

26. Karthikeyan M. "Minimization of moisture reabsorption in dried coal samples". *Drying Technology*, 2008, 26(7): 948–955.

27. Zhang K, You CF. "Effect of upgraded lignite product water content on the propensity for spontaneous ignition". *Energy & Fuels*, 2012, 27(1): 20–26.

28. Gao XZ, Man CB, Hu SJ, et al. "Theoretical and experimental study on spontaneous ignition of lignite during the drying process in a packed bed". *Energy & Fuels*, 2012, 26(11): 6876–6887.

29. Kadioğlu Y, Varamaz M. "The effect of moisture content and air-drying on spontaneous combustion characteristics of two Turkish lignites". *Fuel*, 2003, 82(13): 1685–1693.

30. Zheng HJ, Zhang SY, Dong JX, et al. "Self-combustion characteristics of lignite after dried by high temperature flue gases". *Journal of Engineering for Thermal Energy & Power*, 2015, 30(5): 750–755 [in Chinese].
31. Fei Y, Aziz AA, Nasir S, et al. "The spontaneous combustion behavior of some low rank coals and a range of dried products". *Fuel*, 2009, 88(9): 1650–1655.
32. Solomon PR, Hamblen DG, Carangelo RM, et al. "General model of coal devolatilization". *Energy Fuels*, 1987, 2(4): 405–422.
33. Bongers GD, Jackson WR, Woskoboenko F. "Pressurised steam drying of Australian low-rank coals: Part 1. Equilibrium moisture contents". *Fuel Processing Technology*, 1998, 57(1): 41–54.
34. Lv T, Zhang ZY. "Numerical simulation and experimental study on volatiles release during lignite drying process". *Thermal Power Generation*, 2014, 43(9): 65–70 [in Chinese].
35. Guo X, Zhang SY, Dong AX, et al. "Experimental study on devolatilization of lignite in high temperature flue gas drying". *Coal Conversion*, 2014, 37(1): 23–27 [in Chinese].
36. Agraniotis M, Grammelis P, Papapavlou C, et al. "Experimental investigation on the combustion behaviour of pre-dried Greek lignite". *Fuel Processing Technology*, 2009, 90(9): 1071–1079.
37. Agraniotis M, Stamatis D, Grammelis P, et al. "Numerical investigation on the combustion behaviour of pre-dried Greek lignite". *Fuel*, 2009, 88(12): 2385–2391.
38. Ahmed S, Naser J. "Numerical investigation to assess the possibility of utilizing a new type of mechanically thermally dewatered (MTE) coal in existing tangentially-fired furnaces". *Heat and Mass Transfer*, 2011, 47(4): 457–469.
39. Zhao FT, Witt PJ, Schwarz MP, et al. "Combustion of predried brown coal in a tangentially fired furnace under different operating conditions". *Energy & Fuels*, 2012, 26: 1044–1053.
40. Man CB, Zhu X, Gao XZ, et al. "Combustion and pollutant emission characteristics of lignite dried by low temperature air". *Drying Technology*, 2015, 33(5): 616–631.
41. Wang J, Fan WD, Li Y, et al. "The effect of air staged combustion on NOx emissions in dried lignite combustion". *Energy*, 2012, 37(1): 725–736.
42. Hemis M, Bettahar A, Singh CB, et al. "An experimental study of wheat drying in thin layer and mathematical simulation of a fixed-bed convective dryer". *Drying Technology*, 2009, 27(10): 1142–1151.
43. Tian J, Liu B. "Progress and application of Lignite drying technologies". *Coal Chemical Industry*, 2012, 3(1): 1–5 [in Chinese].
44. Guo F, Li DW, Ren WT. "The application of new drum-type lignite drying system". *Clean Coal Technology*, 2010, 16(1): 29–31 [in Chinese].
45. Yu XH, Liao HY, Zhai JP, et al. "Lignite drying with steam tube rotary dryer and its heat transfer characteristics". *Clean Coal Technology*, 2013, 19(1): 52–54 [in Chinese].
46. Hatzilyberis KS, Androutsopoulos GP, Salmas CE. "Indirect thermal drying of lignite: Design aspects of a rotary dryer". *Drying Technology*, 2000, 18(9): 2009–2049.
47. Feng B, Zhou LR, Dou Y, et al. "Study on drying characteristics of Indonesian lignite in the steam tube rotary dryer". *Guangzhou Chemical Industry*, 2012, 40(9): 79–82 [in Chinese].
48. Pang S, Xu Q. "Drying of woody biomass for bioenergy using packed moving bed dryer: Mathematical modeling and optimization". *Drying Technology*, 2010, 28(5): 702–709.
49. Zhang K, You CF. "Numerical simulation of lignite drying in a packed moving bed dryer". *Fuel Processing Technology*, 2013, 110(2): 122–132.
50. Potter OE, Guang LX, Georgakopoulos S, et al. "Some design aspects of steam-fluidized heated dryers" . *Drying Technology*, 1990, 8(1): 25–39.

51. Jeon DM, Kang TJ, Kim HT, et al. "Investigation of drying characteristics of low rank coal of bubbling fluidization through experiment using lab scale". *Science China Technological Sciences*, 2011, 54(7): 1680–1683.

52. Wang WC. "Laboratory investigation of drying process of Illinois coals". *Powder Technology*, 2012, 225(7): 72–85.

53. Kim HS, Matsushita Y, Oomori M, et al. "Fluidized bed drying of Loy Yang brown coal with variation of temperature, relative humidity, fluidization velocity and formulation of its drying rate". *Fuel*, 2013, 105: 415–424.

54. Klutz HJ, Moser C, Bargen N. "The RWE power WTA process (Fluidized bed drying) as a key for higher efficiency". *Górnictwo i Geoinżynieria*, 2011, 35: 147–153.

55. Hoehne O, Lechner S, Schreiber M, et al. "Drying of lignite in a pressurized steam fluidized bed – Theory and experiments". *Drying Technology*, 2009, 28(1): 5–19.

56. Dong NS. "Techno-economics of modern pre-drying technologies for lignite-fired power plants". IEA Clean Coal Centre, 2014.

57. Lechner S, Hoehne O, Krautz HJ. "Pressurized steam fluidized bed drying (PSFBD) of lignite: Constructional and process optimization at the BTU test facility and experimental results". *Proceedings of the XII Polish Drying Symposium*, Lodz, 2009.

58. Sarunac, N. Power "101: Flue Gas Heat Recovery in Power Plants, Part II". 2010. Available on: http://www.powermag.com/power-101-flue-gas-heat-recovery-in-power-plants-part-ii/ [accessed in July 2018].

59. Aziz M, Kansha Y, Tsutsumi A. "Self-heat recuperative fluidized bed drying of brown coal". *Chemical Engineering & Processing: Process Intensification*, 2011, 50(9): 944–951.

60. Aziz M, Kansha Y, Kishimoto A, et al. "Advanced energy saving in low rank coal drying based on self-heat recuperation technology". *Fuel Processing Technology*, 2012, 104: 16–22.

61. Berghel J, Renström R. "Basic design criteria and corresponding results performance of a pilot-scale fluidized superheated atmospheric condition steam dryer". *Biomass and Bioenergy*, 2002, 23(2): 103–112.

62. Limtrakul S, Rotjanavijit W, Vatanatham T. "Lagrangian modeling and simulation of effect of vibration on cohesive particle movement in a fluidized bed". *Chemical Engineering Science*, 2007, 62(1–2): 232–245.

63. Costa SS, Moris VAS, Rocha SCS. "Influence of process variables on granulation of microcrystalline cellulose in vibro fluidized bed" . *Powder Technology*, 2011, 207(1–3): 454–460.

64. Graham J. "Microwaves for coal quality improvement: the Drycol® project". 2006. Available on: http://www.drycol.com/downloads/Drycol%20Paper%20ACPS-1%2006 0608.pdf [accessed in December 2018].

65. Pusat S, Akkoyunlu MT, Erdem HH. "Evaporative drying of low-rank coal". In *Sustainable Drying Technologies*, Olvera, J.D.R. (Ed.). Intech, 2016, 59–78.

66. Tahmasebi A, Yu J, Li X, et al. "Experimental study on microwave drying of Chinese and Indonesian low-rank coals". *Fuel Processing Technology*, 2011, 92(10): 1821–1829.

67. Tahmasebi A, Yu JL, Han YN, et al. "A kinetic study of microwave and fluidized-bed drying of a Chinese lignite". *Chemical Engineering Research & Design*, 2014, 92(1): 54–65.

68. Song ZL, Yao LL, Jing CM, et al. "Drying behavior of lignite under microwave heating". *Drying Technology*, 2017, 35(4): 433–443.

69. Choicharoen K, Devahastin S, Soponronnarit S. "Performance and energy consumption of an impinging stream dryer for high-moisture particulate materials". *Drying Technology*, 2009, 28(1): 20–29.

70. Waje SS, Thorat BN, Mujumdar AS. "An experimental study of the thermal performance of a screw conveyor dryer". *Drying Technology*, 2006, 24(3): 293–301.

71. Mursito AT, Hirajima T, Sasaki K. "Upgrading and dewatering of raw tropical peat by hydrothermal treatment". *Fuel*, 2010, 89(3): 635–641.
72. Favas G, Jackson WR. "Hydrothermal dewatering of lower rank coals. 1. Effects of process conditions on the properties of dried product". *Fuel*, 2003, 82(1): 53–57.
73. Favas G, Jackson WR. "Hydrothermal dewatering of lower rank coals. 2. Effects of coal characteristics for a range of Australian and international coals". *Fuel*, 2003, 82(1): 59–69.
74. Racovalis L, Hobday MD, Hodges S. "Effect of processing conditions on organics in wastewater from hydrothermal dewatering of low-rank coal". *Fuel*, 2002, 81(10): 1369–1378.
75. Sakaguchi M, Laursen K, Nakagawa H, et al. "Hydrothermal upgrading of Loy Yang brown coal-effect of upgrading conditions on the characteristics of the products". *Fuel Processing Technology*, 2008, 89(4): 391–396.
76. Yu YJ, Liu JZ, Wang RK, et al. "Effect of hydrothermal dewatering on the slurry ability of brown coals". *Energy Conversion and Management*, 2012, 57(2): 8–12.
77. Ge LC, Zhang YW, Xu C, et al. "Influence of the hydrothermal dewatering on the combustion characteristics of Chinese low-rank coals". *Applied Thermal Engineering*, 2015, 90(6): 174–181.
78. Li CZ. *Advances in the Science of Victorian Brown Coal*. Elsevier Science, Amsterdam, 2004.
79. Strauss K. *Method and Device for Reducing the Water Content of Water-Containing Brown Coal*. Germany, EP 0784660B1, 1996.
80. Bergins C. *Science and Technology of the Mechanical Thermal Dewatering*. PhD thesis, Habilitation, Technical University Dortmund, Germany, 2005.
81. Bergins C. "Mechanical/thermal dewatering of lignite. Part 2: A rheological model for consolidation and creep process". *Fuel*, 2004, 83(3): 267–276.
82. Bergins C. "Kinetics and mechanism during mechanical/thermal dewatering of lignite". *Fuel*, 2003, 82(4): 355–364.
83. Bergins C, Hulston J, Strauss K, et al. "Mechanical/thermal dewatering of lignite. Part 3: Physical properties and pore structure of MTE product coals". *Fuel*, 2007, 86(1): 3–16.
84. Wheeler RA, Hoadley AFA, Clayton SA. "Modeling the mechanical thermal expression behaviour of lignite". *Fuel*, 2009, 88(9): 1741–1751.
85. Butler CJ, Green AM, Chaffee AL. "MTE water remediation using Loy Yang brown coal as a filter bed adsorbent". *Fuel*, 2008, 87(6): 894–904.
86. Miura K, Mae K, Ashida R, et al. "Dewatering of coal through solvent extraction". *Fuel*, 2002, 81(11–12): 1417–1422.
87. Fujitsuka H, Ashida R, Miura K. "Upgrading and dewatering of low rank coals through solvent treatment at around 350°C and low temperature oxygen reactivity of the treated coals". *Fuel*, 2013, 114: 16–20.
88. Kanda H, Makino H. "Energy-efficient coal dewatering using liquefied dimethylether". *Fuel*, 2010, 89(8): 2104–2109.
89. Iwai Y, Koujina Y, Arai Y, et al. "Low temperature drying of low rank coal by supercritical carbon dioxide with methanol as entrainer". *Journal of Supercritical Fluids*, 2002, 23(3): 251–255.

8 Energetic and Exergetic Analyses of Coal and Biomass Drying

Junjie Yan and Ming Liu

CONTENTS

8.1 INTRODUCTION

Low-rank coal, including lignite and low-rank bituminous coal, is one of the most widely used fossil fuels. Biomass is an important type of renewable energy. These fuels have the same characteristic of high moisture content. For example, the mass fraction of water in as-received biomass is sometimes as high as 60%.[1] Drying is a proven method to improve the utilization efficiency of high moisture fuels,[2] and can also broaden the utilization areas of these fuels. Various types of dryers that could be applied to dry coal and biomass include evaporative and non-evaporative drying technologies.

Most of the evaporative drying technologies are mature and have been applied in several industrial sectors. However, evaporative drying is an energy intensive process. The heat consumption rate is usually used to evaluate the performance of dryers, which is defined as the amount of energy consumed to evaporate 1 kg moisture (water in coal and biomass). However, the drying medium used to dry lignite can be at different temperatures and in different forms, thus the energy possessed by the drying medium can be in different grades. In this regard, the heat consumption rate could not perfectly reflect performances of dryers based on the second law of thermodynamics. Exergy is a property of thermodynamics expressing the maximum useful work contained in energy medium.[3] It is widely used to evaluate the efficiency of the performance of energy processes based on the grade of energy.

Therefore, models for energetic and exergetic analyses of dryers are provided in this chapter, and can be used to analyze the energy and exergy consumption rates of dryers. Moreover, results of energy and exergetic analyses can guide the selection, design, and optimization of drying technologies.

Most of the low-rank coal is consumed in thermal power plants. Biomass can be used separately or blended with coal in thermal power plants. The integration of drying operations within power plants is very attractive for the application of drying technologies, which can significantly increase the efficiency of power plants. For the same quantity of energy, if a drying medium with low exergy is used to dry the low-rank coal and biomass, the overall performance of utilization of these fuels is significantly improved. Therefore, quantifying the influence of fuel drying on the power plant efficiency is very important for the application of dryers. Many researchers have analyzed the influence of drying on coal-fired power plants[4-9] and biomass-fired power plants.[10-13] In this chapter, models to simulate the integration of dryers with power plants are also introduced, which will help readers to investigate the quantitative results for the influence of dryers on power plant efficiency.

Then, the dryers' energy and exergy consumption rates are analyzed with some typical dryers as examples. Finally, influences of drying on biomass- and lignite-fired power plants are evaluated with case studies. In these studies, both flue-gas and steam are used to dry raw solid fuel.

8.2 MODELS ON ENERGETIC AND EXERGETIC ANALYSES OF DRYERS

8.2.1 ENERGY BALANCE MODEL

Drying could be categorized as evaporative drying and non-evaporative drying. Most of the evaporative drying technologies are mature and industrially applicable.

However, evaporative drying is an energy intensive process, because extensive heat is needed for the water in the solid fuel to evaporate. For evaporative drying technologies, steam, air, or flue-gas can be used as a drying heat source. In hot air drying and flue-gas drying, the hot and humid exhaust drying medium induces significant heat loss, thus reducing energy efficiency.

The drying process could be expressed as shown in Figure 8.1, in which 1 kg raw solid fuel (as received, wet) is fed into the dryer, and λ kg water is dried out. m_h kg heating medium (flue-gas or steam) releases heat in the dryer to dry the raw solid fuel. In hot air drying, m_a kg air is used for drying 1 kg raw solid fuel. In some dryers, the heating medium comes directly into contact with wet solid fuel and leaves the dryer as the dryer exhaust, while in some indirect dryers (rotary-tube dryer, etc.), the heating medium provides heat for drying in the dryer.

The degree of drying, λ, is defined as the mass of water removed from per unit mass of raw solid fuel as follows:

$$\lambda = \frac{M_{raw} - M_{upg}}{100 - M_{upg}} \tag{8.1}$$

where M_{raw} and M_{upg} are the mass ratio of water contained in per unit mass of raw and dried solid fuel, respectively, in %.

The minimum energy consumption, which is only absorbed by solid fuel to increase temperature and cause water to evaporate, for drying 1 kg raw solid fuel could be evaluated with

$$q_{d0} = \left[\lambda \cdot (h_{dw} - h_{w0}) + (1 - \lambda) \cdot (h_{c1} - h_{c0}) \right] \times 1\text{kg} \tag{8.2}$$

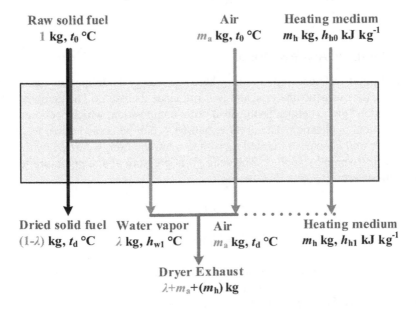

FIGURE 8.1   Schematic diagram of dryer.

where h_{dw} and h_{w0} are enthalpies of the water contained in the dryer's drying medium and raw solid fuel, respectively, in kJ kg^{-1}; h_{c1} and h_{c0} are enthalpies of dried solid fuel at the outlet and inlet positions, respectively, in kJ kg^{-1}.

The energy possessed by exhaust drying air equivalent to 1 kg raw solid fuel is:

$$q_{da} = m_a \cdot Cp_a (t_d - t_0) \tag{8.3}$$

where m_a is the mass of air, kg; Cp_a is the specific heat capacity of air, kJ kg^{-1} K^{-1}; t_d and t_0 are temperatures of the dryer exhaust and ambient respectively, °C.

Based on the energy balance in the dryer, the mass of heating medium for drying of 1 kg raw solid fuel is

$$m_h = \frac{q_{d0} + q_{da}}{\eta_d \cdot (h_{h0} - h_{h1})} \tag{8.4}$$

where h_{h0} and h_{h1} are enthalpies of heat medium at inlet and outlet positions in the dryer, kJ kg^{-1}; η_d is the thermal efficiency of the dryer.

The energy consumption rate to remove 1 kg water from raw solid fuel is calculated by:

$$q_{H_2O} = \frac{q_{d0} + q_{da}}{\eta_d \cdot \lambda} \tag{8.5}$$

To evaluate the mass flow rate of drying heat medium, the mass flow rate of drying medium needed to dry out 1 kg water is defined as

$$K_h = \frac{m_h}{\lambda} \tag{8.6}$$

8.2.2 Exergetic Analysis Model

The dead point (reference conditions where the exergy is defined as zero) is the benchmark for exergetic analysis, and will influence its results. The composition of saturated moist gas is defined as the dead point composition, which is chosen based on its practical application. The dryer exhaust gas can be cooled down to ambient temperature and becomes saturated moist gas when the exergy carried by the dryer exhaust is recovered by cooling. The dead point pressure and temperature for exergetic analysis are

$$p_0 = 0.1 \, \text{MPa} \tag{8.7}$$

$$T_0 = 293.15 \, \text{K} \tag{8.8}$$

In the drying process, substances include solid fuel, gas (flue-gas and air), water (liquid and steam), and dryer exhaust. Assumptions of the ideal gas mixture for the dryer exhaust are used. The exergy of the water component is calculated with

$$E_w = h_w - h_{w0} - T_0\left(s_w - s_{w0}\right) + \bar{R}_w T_0 \ln \frac{y_w}{y_{w0}} \tag{8.9}$$

where h_w and h_{w0} are the enthalpies of water at given conditions and dead point conditions respectively, kJ kg^{-1}; s_w and s_{w0} are the entropies at given conditions and dead point conditions respectively, kJ kg^{-1} K^{-1}; $\bar{R}_w$ is the gas constant of water vapor, kJ kg^{-1} K^{-1}; y_w and y_{w0} are volume fractions of water vapor at given conditions and dead point conditions.

It is assumed that flue-gas or air component has constant specific heat (average specific heat Cp$_g$, kJ kg^{-1} K^{-1}). Then the exergy of the gas component is

$$e_g = Cp_g\left(T_g - T_0 - T_0\ln\frac{T_g}{T_0}\right) + R_g T_0 \ln \frac{y_g}{y_{g0}} \tag{8.10}$$

where T_g is the temperature at given condition, K; $\bar{R}_g$ is the gas constant of gas, kJ kg^{-1} K^{-1}; y_g and y_{g0} are volume fractions of gas at given conditions and dead point conditions.

The physic exergy carried by the dried solid fuel is

$$e_l = Cp_l\left(T_d - T_0 - T_0\ln\frac{T_d}{T_0}\right) \tag{8.11}$$

where Cp$_l$ is the constant specific heat of solid fuel, kJ kg^{-1} K^{-1}; T_d is the temperature of solid fuel at the outlet of dryer, K.

To quantitatively compare the thermodynamic performance of various dryers, the exergy feeding rate per 1 kg water removed from the solid fuel feed is defined as

$$\xi_f = \frac{E_h}{\lambda} \tag{8.12}$$

where E_h is the exergy of drying medium fed into the dryer, kJ.

The exergy carried by the dried solid fuel or dryer exhaust can be recovered by heat recovery, whereas the internal exergy loss due to irreversible processes in the dryer and the external exergy loss along with heat loss of the dryer could not be recovered. The exergy consumption rate is defined as

$$\xi_c = \frac{E_h - E_l - E_e}{\lambda} \tag{8.13}$$

From the above discussion, the exergy feeding rate ξ_f (kJ (kg H$_2$0)$^{-1}$) expresses the exergy feeding rate to dry 1 kg water out from raw solid fuel, and the exergy consumption rate ξ_c (kJ (kg H$_2$0)$^{-1}$) expresses the irreversible losses of dryers.

8.3   MODELS TO ANALYZE THE INTEGRATION OF DRYERS WITH POWER PLANTS

To analyze the influence of wet solid fuel drying on the performance of power plants, an overall energy balance calculation for power plants has been carried out and presented in this section. A conventional power plant (CPP) without the integration of solid fuel dryer is the benchmark for performance evaluation. In this section, the main models to analyze the integration of dryers with power plants are developed. The integration of solid fuel dryers will increase the heating values of the fuel, thus improving the efficiency of the power plant. Based on these results, the efficiency of the power plant could be determined and assessed.

In the following analyses, three types of model will be presented and used. These include (1) models for calculation of heating values and exergies of the raw and dried fuel; (2) models for boiler heat balances which are used to obtain the boiler thermal efficiency of the power plant integrated with dryer (PPD) based on the benchmark parameters of the boiler in CPP, and (3) models of steam turbine system which are used to calculate the performance of the steam turbines.

8.3.1 Models for Heating Values and Exergies of the Solid Fuels

The heating value is the amount of heat released by fuel during combustion, which can be expressed either as the higher heating value (HHV) or as the lower heating value (LHV). During the drying process of wet solid fuel (coal or biomass), vaporization of water is a physical process. In this process, the combustible matters in the solid fuel don't change, thus the higher heating value of the dried solid fuel can be obtained with the following equation:

$$\text{HHV}_{\text{upg}} = \frac{100 - M_{\text{upg}}}{100 - M_{\text{raw}}} \text{HHV}_{\text{raw}} \tag{8.14}$$

where HHV_{raw} is the higher heating value of solid fuel on the as-received basis.

The lower heating value of dried solid fuel could then be calculated with

$$\text{LHV}_{\text{upg}} = \text{HHV}_{\text{upg}} - 206 H_{\text{upg}} - 23 M_{\text{upg}} \tag{8.15}$$

where H_{upg} is the mass ratio of hydrogen contained in dried solid fuel, %.

Based on thermodynamics, the exergy of fuel could be expressed as

$$e_{\text{fuel}} = \text{HHV}_{\text{fuel}} \tag{8.16}$$

8.3.2 Boiler Heat Balance Models

When dryers are integrated with power plants, the temperature of boiler exhaust flue-gas can be decreased due to the reduced humidity with dried fuel, therefore, the boiler thermal efficiency can be increased as a result. Other heat losses, except for the heat loss along with the boiler exhaust flue-gas, are assumed to be the same for the boiler of PPD compared with the boiler of CPP. The diagrams of boiler heat balance of CPP, PPDF, and PPDS are shown in Figures 8.2, 8.3, and 8.4, respectively. The meanings of symbols in these figures are provided in Table 8.1.

The general energy balance equation of the boiler for CPP can be expressed as

$$\dot{Q}_r^0 = \dot{Q}_1^0 + \dot{Q}_2^0 + \dot{Q}_3^0 + \dot{Q}_4^0 + \dot{Q}_5^0 + \dot{Q}_6^0 \tag{8.17}$$

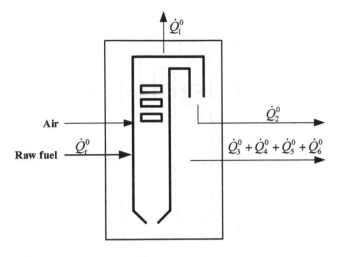

FIGURE 8.2   Boiler heat balance of CPP.

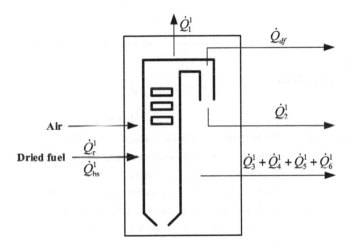

FIGURE 8.3   Boiler heat balance of PPDF.

$$\dot{Q}_r^0 = \dot{m}_f \cdot \mathrm{LHV}_{\mathrm{raw}} \qquad (8.18)$$

where $\dot{m}_f$ is the mass flow rate of feeding fuel; $\mathrm{LHV}_{\mathrm{raw}}$ is the LHV of the as-received fuel.

The boiler thermal efficiency of CPP could be expressed as

$$\eta_{b0}^{\mathrm{en}} = q_1^0 = 1 - q_2^0 + q_3^0 + q_4^0 + q_5^0 + q_6^0 \qquad (8.19)$$

8.3.2.1   Using Flue-Gas as Heat Resource

Unlike the CPP, the PPDF uses dried fuel as the feeding fuel. Some sensible heat of $\dot{Q}_{fs}^1$ is input to the boiler along with the dried fuel because the temperature of the

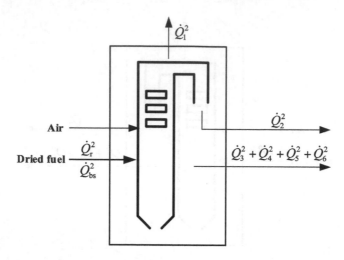

FIGURE 8.4 Boiler heat balance of PPDS.

TABLE 8.1
Meanings of Symbols

CPP	PPDF	PPDS	Meanings
$\dot{Q}_r^0$	$\dot{Q}_r^1$	$\dot{Q}_r^2$	The heat entering the boiler envelope (input heat)
$\dot{Q}_1^0$	$\dot{Q}_1^1$	$\dot{Q}_1^2$	The heat available in the boiler (output heat to steam turbines)
$\dot{Q}_2^0$	$\dot{Q}_2^1$	$\dot{Q}_2^2$	The heat loss caused by the boiler exhaust gases
$\dot{Q}_3^0$	$\dot{Q}_3^1$	$\dot{Q}_3^2$	The heat loss caused by unburned gaseous combustibles
$\dot{Q}_4^0$	$\dot{Q}_4^1$	$\dot{Q}_4^2$	The heat loss caused by unburned solid combustibles
$\dot{Q}_5^0$	$\dot{Q}_5^1$	$\dot{Q}_5^2$	The heat loss caused by radiation and convection
$\dot{Q}_6^0$	$\dot{Q}_6^1$	$\dot{Q}_6^2$	The other heat losses
\	$\dot{Q}_{df}$	\	The energy loss along with dryer exhaust
\	$\dot{Q}_{fs}^1$	$\dot{Q}_{fs}^2$	The sensible heat input to the boiler with the dried fuel because of temperature increase

dried fuel is higher than the environmental temperature. Moreover, some energy of $\dot{Q}_{df}$ is led to the dryer, which is an energy loss for the dryer. Then, the boiler energy balance equation can be established as:

$$\dot{Q}_r^1 + \dot{Q}_{fs}^1 = \dot{Q}_1^1 + \dot{Q}_2^1 + \dot{Q}_3^1 + \dot{Q}_4^1 + \dot{Q}_5^1 + \dot{Q}_6^1 + \dot{Q}_{df} \qquad (8.20)$$

$$\dot{Q}_r^1 = \dot{m}_{f1} \cdot \text{LHV}_{\text{upg}} \qquad (8.21)$$

where LHV_{upg} is the LHV of the dried fuel.

Rewriting the above equation on an input basis gives

$$q_1^1 + q_3^1 + q_4^2 + q_5^1 + q_6^1 + q_{df}^1 - q_{fs}^1 = 1 \tag{8.22}$$

Boiler efficiency can then be expressed as

$$\eta_{b1}^{en} = q_1^1 = 1 - q_{df}^1 - (q_3^1 + q_4^1 + q_5^1 + q_6^1) + q_{fs}^1 \tag{8.23}$$

It can be assumed that heat losses excluding the heat loss from exhaust flue-gas in boilers with integration with solid fuel drying are the same as the heat losses in boilers without integration with solid fuel drying, therefore,

$$q_i^1 = q_i^0 \left(i = 3 \sim 6 \right) \tag{8.24}$$

The sensible heat of the feeding fuel is

$$q_{fs}^1 = \frac{c_{b1} \cdot (t_{b1} - t_0)}{LHV_{upg}} \tag{8.25}$$

The rate of energy led to the dryer (q_{df}^1) could be obtained with the aid of dryer models.

8.3.2.2 Using Steam Extraction as Heat Resource

Similarly to the boiler of the PPDF, some sensible heat of Q_{fs}^2 is input to the boiler of the PPDS with dried fuel. Then, the energy balance equation of the boiler for PPDS can be expressed as:

$$\dot{Q}_r^2 + \dot{Q}_{bs}^2 = \dot{Q}_1^2 + \dot{Q}_2^2 + \dot{Q}_3^2 + \dot{Q}_4^2 + \dot{Q}_5^2 + \dot{Q}_6^2 \tag{8.26}$$

$$\dot{Q}_r^2 = \dot{m}_{f2} \cdot LHV_{upg} \tag{8.27}$$

Rewriting the above equation as ratios to the $\dot{m}_{f2} \cdot LHV_{upg}$ gives

$$q_1^2 + q_2^2 + q_3^2 + q_4^2 + q_5^2 + q_6^2 - q_{fs}^2 = 1 \tag{8.28}$$

The boiler efficiency could be expressed as

$$\eta_{b2}^{en} = q_1^2 = 1 - q_2^2 - \left(q_3^2 + q_4^2 + q_5^2 + q_6^2 \right) + q_{fs}^2 \tag{8.29}$$

Heat losses from boilers with integration with solid fuel drying are again assumed to be the same as the heat losses in boilers without integration with solid fuel drying, except for the heat loss along with the boiler exhaust:

$$q_i^2 = q_i^0 \left(i = 3 \sim 6 \right) \tag{8.30}$$

The sensible heat of the feeding fuel (q_{fs}^2) could be calculated with Equation (8.25), and the heat loss caused by the boiler exhaust gases (q_2^2) could be calculated with the components and temperature of the boiler flue-gas.

8.3.3 MODELS TO ANALYZE THE ENERGY BALANCE OF THE STEAM TURBINE SYSTEM

Solid fuels may be dried in a rotary-tube dryer or WTA dryer using steam as the heating medium, which can be extracted from the steam turbine system to the dryer. This arrangement will certainly influence the efficiency of the steam turbine system. The following assumptions are made for the thermodynamic analysis of steam turbine systems between PPD and CPP: (1) the parameters of live steam are the same; (2) the isentropic efficiencies of the steam turbines are the same; (3) the terminal temperature differences (upper and lower terminal temperature differences) for the regenerative heaters are the same. It should be stated that these assumptions are often applied in thermal power plant analyses.

Therefore, the steam turbine parameters of the CPP can be directly used for PPDF. However, for PPDS, some steam with a mass flow rate of $\dot{D}_1$ is extracted from the steam turbine system to the dryer:

$$\dot{D}_1 = \frac{\dot{Q}_d}{h_{h0} - h_{h1}}$$
(8.31)

where h_{h0} and h_{h1} are the enthalpies of the steam to and from the dryer, respectively, kJ/kg. $\dot{Q}_d$ is the heat load of the dryer, which could be calculated with models in Section 8.1. The condensate of steam extraction to the dryer is normally sent back to a deaerator of the steam turbine system. With the assumption of steam turbine system, the water flow rate outflowing the deaerator is the same as that of CPP; therefore, the water flow rate entering deaerator and heaters before the deaerator is reduced by $\dot{D}_1$.

For the deaerator, which is No. k heater in the heater train of heat regenerative system, as shown in Figure 8.5, applying an energy balance for the deaerator, we can get the following equation,

$$D_1 \cdot \left(h_{wk} - h_{w(k-1)}\right) + D_1 \cdot \left(h_{h1} - h_{wk}\right) = \Delta D_k \left(h_k - h_{wk}\right)$$
(8.32)

where ΔD_k expresses the steam variation of the deaerator, between PPDS and CPP systems.

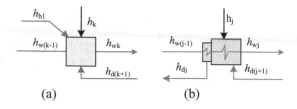

(a)   (b)

FIGURE 8.5   Regenerative heaters. (a) Deaerator (No.k), (b) Heater (No.j).

From Equation (8.32), we can get steam extraction variation of deaerator, ΔD_k, as follows,

$$\Delta D_k = \frac{D_1 \cdot \left(h_{h1} - h_{w(k-1)}\right)}{h_k - h_{wk}} \tag{8.33}$$

The heaters before the deaerator are always shell and tube heat exchangers in a modern power plant. Parameters of No. j heater are shown in Figure 8.5b, including feeding water enthalpy $h_{w(j-1)}$, export water enthalpy h_{wj}, steam extraction enthalpy h_j, drain water enthalpy h_{dj} and drain water enthalpy from No. $(j+1)$ heater, $h_{d(j+1)}$. In comparison with heat regenerative system without fuel drying, the steam extraction variation is ΔD_j, and drain water from previous higher-pressure heater (No. $(j+1)$) variation is ΔD_{dj} for the No. j heater.

We assume that terminal temperature differences (upper and lower temperature differences) of heater keeps unchanged, so the steam extraction variation can be calculated with parameters of the heat regenerative system of CPP as in the following equation,

$$\Delta D_j = \frac{D_1\left(h_{wj} - h_{w(j-1)}\right) \cdot -\Delta D_{d(j+1)} \cdot \left(h_{d(j+1)} - h_{dj}\right)}{h_j - h_{dj}} \tag{8.34}$$

and the drain water of No. $(j+1)$ heater variation $\Delta D_{d(j+1)}$ can be calculated by the following equation,

$$\Delta D_{d(j+1)} = \sum_{i=j+1}^{k-1} \Delta D_i \tag{8.35}$$

If this part of steam had expanded in the turbines then it could produce an extra work ΔW_j, which can be expressed as follows,

$$\Delta W_j = \Delta D_j \cdot \left(h_j - h_n\right) \tag{8.36}$$

Therefore, the work variation of steam ΔW between CPP and PPDS can be expressed as follows,

$$\Delta W = D_1 \cdot \left(h_f - h_n\right) - \sum_{j=1}^{k} \Delta W_j \tag{8.37}$$

Then the reduction of system power output can be expressed by the following equation,

$$\Delta W_e = \Delta W \cdot \eta_m \cdot \eta_e \tag{8.38}$$

where η_m is the mechanical efficiency of the steam turbine; η_e is the generator efficiency.

8.4   DRYERS' ENERGY AND EXERGY CONSUMPTION RATES

As a case study, the energy and exergy consumption rates of evaporative drying technologies will be calculated in this section. Yimin lignite is used as the
feeding fuel which composition is shown in Table 8.2. The moisture content of
raw lignite is 39.5% and it is assumed to be dried to 15%. The constant specific
heat of lignite is 1.3 kJ kg^{-1} K^{-1}. Dryers using steam as heating a source are
classified as steam dryers, including the rotary-tube dryer and WTA dryer. The
characteristics and operation parameters of steam dryers are listed in Table 8.3.
When flue-gas is used to dry lignite, the dryers are classified as flue-gas dryers. Characteristics and operation conditions of flue-gas dryers are listed in
Table 8.4.

8.4.1   ENERGY ANALYSIS OF DRYERS

The energy balance equations are applied for the lignite dryers listed in Tables 8.3
and 8.4 based on the mass flow rate of a heat source for drying 1 kg water out of lignite, and results are shown in Figure 8.6. As shown in Figure 8.6, the mass flow rate
of heat source varies significantly. The heat released by per unit mass of flue-gas is
significantly lower than that released by per unit mass of steam. Therefore, K_h values
for flue-gas dryers are greater than that for steam dryers. The K_h is above 29 for the
moving bed dryer, because the temperature drop of flue-gas through the moving bed
dryer is only 95°C. For the WTA dryer and rotary-tube dryer, the rates of drying
medium are 1.38 and 1.49, respectively.

TABLE 8.2
Compositions of Yimin Lignite

C(wt%)	H(wt%)	S(wt%)	O(wt%)	N(wt%)	Ash(wt%)	H$_2$O(wt%)	LHV(MJ/kg)
34.59	2.03	0.14	11.3	0.35	12.09	39.5	11.79

TABLE 8.3
Operation Parameters of Steam Dryers[14]

Dryer Type	Characteristics	Heating Medium Parameters	
		Inlet	Outlet
Rotary-tube dryers	Using air as carrier of evaporative moisture; consist of a drum equipped with tubes.	~180°C/0.4–0.5 MPa	~Saturated water
WTA dryers	Lignite drying in slightly superheated steam; steam fluid-bed with internal heaters.	~140°C/0.32 MPa	~Saturated water

TABLE 8.4
Operation Parameters of Flue-Gas Dryers[15]

Dryer Type	Characteristics	Heating Medium Temperature/°C	
		Inlet	Outlet
Rotary	Drying along with disintegration; co-current mode.	750	120
Pneumatic	Short drying time; lignite lifted by drying gas during pneumatic transport drying.	600	100
Fluid-bed	Easy to control; High drying intensity due to good mixing and high temperature heating medium.	450	75
Moving bed	Possibility of full automation; compact construction and simple design.	175	80

8.4.2 COMPARISON OF EXERGY FEEDING AND CONSUMPTION RATES FOR DIFFERENT DRYERS

Based on the heat balance of lignite dryers, exergetic analyses were conducted. The exergy feeding and consumption rates for lignite dryers were compared in Figure 8.7, which shows that the exergy feeding rate and exergy consumption rate for different dryers vary greatly. In the rotary dryer, the flue-gas at a temperature of 750°C is used to dry lignite, so the exergy feeding rate is as high as 1837.9 kJ (kg H_2O)$^{-1}$. If the physical exergy contained in dryer exhaust and dried lignite could be recovered, the energy

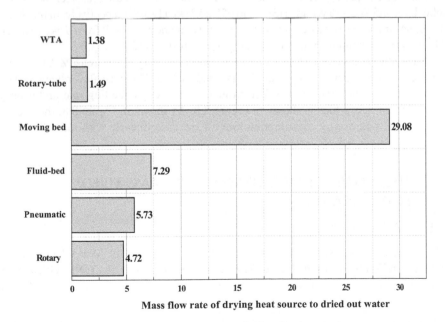

FIGURE 8.6 Comparison on mass flow rate of drying heat source.

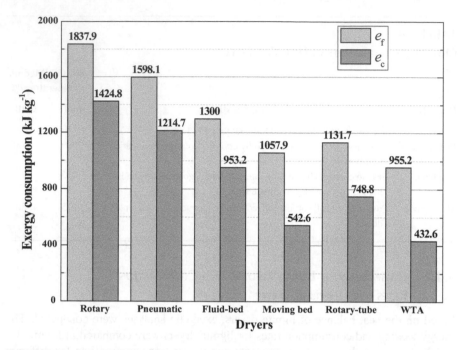

FIGURE 8.7   Comparison of exergy consumption rate.

consumption rate could be decreased to 1424.8 kJ (kg H_2O)$^{-1}$. A low temperature heat source is used to dry lignite for the steam dryers. The exergy feeding rates for the rotary-tube dryer and WTA dryer are 1131.7 and 955.2 kJ (kg H_2O)$^{-1}$, respectively. In the WTA dryer, no air is used as carrier gas, so more exergy can be recovered from dryer exhaust. The exergy consumption rate for WTA dryer is only 432.6 kJ (kg H_2O)$^{-1}$.

Air is always used as carrier gas for the rotary-tube dryer. In Figure 8.7, the mass flow rate of air for drying 1 kg water is assumed as 3 kg. The mass flow rate of carrier gas for rotary-tube dryer will indeed influence the exergy feeding rate and exergy consumption rate, which is shown in Figure 8.8. As shown in Figure 8.8, the exergy feeding rate and exergy consumption rate all increase linearly with mass flow rate of carrier air.

8.5   THE INFLUENCE OF DRYING ON BIOMASS-FIRED POWER PLANTS*

8.5.1   A CASE STUDY

The reference case used as a benchmark in this work is a 12 MW CPP which is depicted in Figure 8.9. In the figure, APH, FSH, FP, DA, RH, CP, COND, GS, and PS are used to denote air pre-heater, flue-gas-steam heat exchangers in a boiler, feedwater

---

* Some contents in this section have been published in authors' paper (https://doi.org/10.1016/j.applther maleng.2017.08.156)

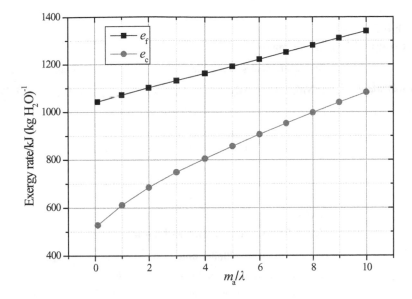

FIGURE 8.8 Influence of carrier air on exergy feeding and consumption rates for rotary-tube dryer.

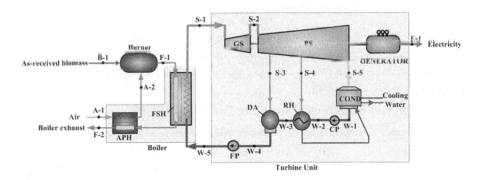

FIGURE 8.9 Conventional biomass power plant.

pump, deaerator, regenerative heater, condensation water pump, condenser, governing stage of steam turbine, and pressure stage of steam turbine, respectively. The CPP is divided into three subsystems, namely, burner, boiler, and turbine unit. The compositions of the feeding fuel of the CPP are shown in Table 8.5, and parameters of the CPP main units are shown in Table 8.6.

The thermodynamic properties of the state points marked in Figure 8.9 are calculated and the results are listed in Table 8.7.

The exergetic analysis of the CPP is visualized with a Sankey diagram, which is shown in Figure 8.10. The exergetic efficiency of the CPP is 21.15%. More than 50% exergy is consumed in the burner because of the high moisture content of biomass, while 14.16% and 5.46% exergies are lost in the boiler unit and turbine unit,

TABLE 8.5
The Compositions of Biomass[a]

Biomass Variety	Proximate Analysis on as-Received Basis (%)				Heating Value (MJ kg⁻¹)	
	VM_{ar}	FC_{ar}	M_{ar}	A_{ar}	HHV_{ar}	LHV_{ar}
Forest residue	34.5	7.3	56.8	1.4	8.46	6.68

[a] The compositions of biomass are based on Ref.[1]. VM, volatile matter; FC, fixed carbon; M, moisture content; A, Ash; ar, as-received basis; daf, dry, ash-free basis; HHV, higher heating value; LHV, lower heating value.

TABLE 8.6
Basic Parameters of the Reference Case

Boiler unit	
Thermal efficiency based on LHV_{ar}	86.00%
Thermal efficiency of steam pipe	99.00%
Turbine unit	
Absolute internal efficiency	32.51%
Mechanical efficiency	99.00%
Generator efficiency	97.67%
Power plant	
Energetic efficiency based on LHV_{ar}	26.77%

TABLE 8.7
Thermodynamic Properties of the State Points of the CPP

Point	Substance	Temperature (°C)	Pressure (MPa)	Mass Flow Rate (kg s⁻¹)	Exergy (kW)
B-1	Biomass	20	0.1000	6.9770	59016.0
F-1	Flue-gas	1361.6	0.1000	28.3750	27756.8
F-2	Flue-gas	150.0	0.1000	28.3750	742.2
A-1	Air	0.0	0.1000	21.5031	0.0
A-2	Air	250.0	0.1000	21.5031	1315.0
S-1	Steam	435.0	3.4300	13.8889	17560.8
S-2	Steam	355.0	1.4500	13.8889	14867.0
S-3	Steam	158.2	0.2131	0.3492	227.3
S-4	Steam	94.3	0.0823	1.4169	689.9
S-5	Steam	33.6	0.0052	12.1228	1184.6
W-1	Water	33.6	0.0052	13.5397	16.0
W-2	Water	33.6	0.5390	13.5397	23.4
W-3	Water	89.8	0.5390	13.5397	414.1
W-4	Water	104.3	0.1180	13.8889	595.0
W-5	Water	105.2	5.8800	13.8889	685.5
E-1	Electricity	\	\	\	12483.7

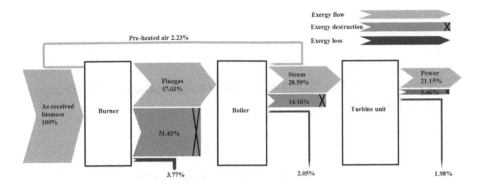

FIGURE 8.10 Sankey diagram of exergetic analysis of the CPP.

respectively. The exergy analysis shows the high potential to reduce the exergy consumption with the method of exergy recuperation,[16, 17] that is, using low-grade energy to dry the biomass.

8.5.2 FLUE-GAS DRYING

The flue-gas dryer, which is a direct rotary dryer, is shown in Figure 8.11. The boiler exhaust flue-gas (F-2 in Figure 8.9) is led to the dryer and used as the drying medium, which comes into contact with the biomass directly. The as-received biomass (B-0) is sent to the dryer for drying. The dried biomass (B-1 in Figure 8.9) is then sent to the burner. The dryer exhaust gas (F-3) is sent to the flue-gas treatment system. The performance data of the flue-gas dryer are listed in Table 8.8.

With the models developed above, the thermodynamic properties of the state points in the PPDF (marked in Figure 8.9 and Figure 8.11) are calculated and shown in Table 8.9. The exergetic analysis of the PPDF is visualized with a Sankey diagram, which is shown in Figure 8.12.

TABLE 8.8
Performance Data of the Flue-Gas Dryer

Drying Medium Type	Flue-Gas
Thermal efficiency[a]	70%
Biomass moisture content out[a]	15%
Drying medium temperature in[a]	250°C
Drying medium temperature out[b]	104°C

[a] The thermal efficiency, biomass moisture content out and drying medium temperature in, is based on Ref.[12].
[b] The drying medium temperature out is based on Ref.[18].

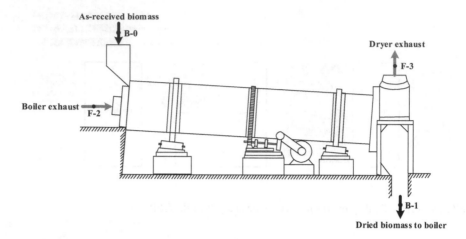

FIGURE 8.11 Flue-gas dryer and its integration with the power plant.

Comparison of Figure 8.10 with Figure 8.12 indicates that the exergetic efficiency of the power plant increases by 0.27% with biomass drying. The above results show:

1. The exergy consumption in the burner decreases by 5.05% because of the integration of the flue-gas dryer.
2. About 3.58% more exergy is consumed in the boiler (heat exchange process in the APH and FSH) because we maintain the parameters of the working fluid to be the same as those of the CPP; however, the temperature of flue-gas increases.
3. Roughly between 1.34 and 1.66% exergies are lost and destructed in the dryer, respectively. Therefore, the exergetic efficiency of the PPDF increases because of the decrease of exergy destruction in the burner. The exergetic efficiency of the PPDF could be further increased.
4. The exergetic efficiency of the heat exchange process in the boiler (η_{bh}^{ex}) could be improved by increasing the temperature and pressure of the working fluid. The exergy destruction in the boiler heat exchange process could be decreased as a result.
5. The exergy loss in the dryer could be decreased by increasing the thermal efficiency of the dryer.

8.5.3 Steam Drying

The steam dryer, such as a steam-tube rotary dryer, is shown in Figure 8.13 in which the steam is extracted from the steam turbine (S-6). The steam releases heat in the dryer and subsequently condenses. The steam condensate (W-6) as saturated water is reused and the heat is recovered in the DA. The as-received biomass (B-0) is dried in the steam dryer, and the dried biomass (B-1) is sent to the burner. The dryer exhaust, which primarily comprises of water vapor, is led to the dryer exhaust treatment system. The performance data of the steam dryer are listed in Table 8.10.

TABLE 8.9

Thermodynamic Properties of the State Points in the PPDF

Point	Substance	Temperature (°C)	Pressure (MPa)	Mass Flow Rate (kg s⁻¹)	Exergy (kW)
B-0	Biomass	20.0	0.1000	6.8886	58267.8
B-1	Dried biomass	104.0	0.1000	5.7228	58267.8
F-1	Flue-gas	1545.1	0.1000	26.8495	30213.0
F-2	Flue-gas	250.0	0.1000	26.8495	1837.4
F-3	Dryer exhaust	104.0	0.1000	28.0153	331.6
A-1	Air	20.0	0.1000	17.6376	0.0
A-2	Air	250.0	0.1000	17.6376	1078.6
S-1	Steam	435.0	3.4300	13.8889	17560.8
S-2	Steam	355.0	1.4500	13.8889	14867.0
S-3	Steam	158.2	0.2131	0.3492	227.3
S-4	Steam	94.3	0.0823	1.4169	689.9
S-5	Steam	33.6	0.0052	12.1228	1184.6
W-1	Water	33.6	0.0052	13.5397	16.0
W-2	Water	33.6	0.5390	13.5397	23.4
W-3	Water	89.9	0.3290	13.5397	411.7
W-4	Water	104.3	0.1180	13.8889	595.0
W-5	Water	105.2	5.8800	13.8889	685.5
E-1	Electricity	\	\	\	12483.7

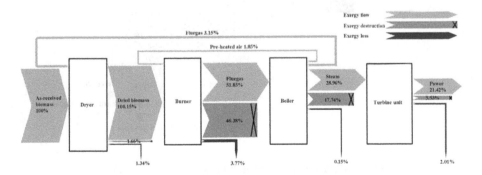

FIGURE 8.12 Sankey diagram of exergetic analysis of the PPDF.

With the models developed above, the thermodynamic properties of the state points in the PPDS (marked in Figure 8.9 and Figure 8.13) are calculated and shown in Table 8.11. The exergetic analysis results of PPDS are visualized with a Sankey diagram, which is shown in Figure 8.14.

As shown in the comparison of Figures 8.10 and 8.14, η_{plant}^{ex} of the PPDS increases only by 0.01% in comparison with that of the CPP. Results show the following.

1. The exergy consumed in the burner decreases by 15.51% because of the integration of the steam dryer.

2. About 9.45% more exergy is consumed in the boiler heat exchange process, because the steam parameters stay the same; however, the flue-gas temperature increases.
3. Roughly between 3.39 and 2.86% exergies are lost and destructed in the dryer.

The exergy loss of the dryer is mainly due to the exergy contained in the dryer exhaust. It could be decreased via waste heat recovery from the dryer exhaust; however, this aspect is beyond the scope of this work and thus is not discussed. The exergy consumption in the dryer is caused by the high temperature difference between the heating medium (steam to dryer) and the biomass. Therefore, η_{plant}^{ex} of the PPDS can be further enhanced if steam with low grade energy is used as the heat source, however, a smaller heat transfer temperature difference will lead to a bigger

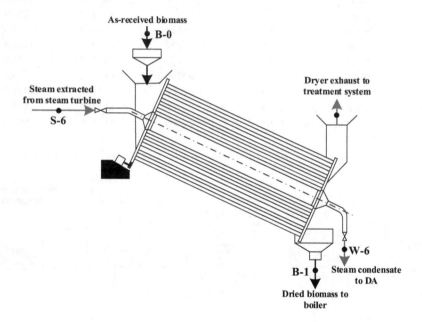

FIGURE 8.13   Steam tube dryer and its integration with the power plant.

TABLE 8.10
Performance Data of the Steam-Tube Rotary Dryer[a]

Drying Medium Type	Steam
Biomass moisture content out	15%
Steam pressure	1100 kPa
Thermal efficiency	85%

[a] The performance data are based on Ref.[12.]

TABLE 8.11

Thermodynamic Properties of the State Points of the PPDS

Point	Substance	Temperature (°C)	Pressure (MPa)	Mass Flow Rate (kg s⁻¹)	Exergy (kW)
B-0	Biomass	20.0	0.1000	6.9749	58997.8
B-1	Dried biomass	100.0	0.1000	3.5449	58997.8
F-1	Flue-gas	1979.6	0.1000	24.9363	36304.3
F-2	Flue-gas	150.0	0.1000	24.9363	597.2
F-3	Dryer exhaust	100.0	0.1000	3.4300	1786.3
A-1	Air	0.0	0.1000	10.9252	0.0
A-2	Air	250.0	0.1000	10.9252	668.1
S-1	Steam	435.0	3.4300	16.9242	21398.6
S-2	Steam	355.0	1.4500	16.9242	18116.1
S-3	Steam	158.2	0.2131	0.2862	186.3
S-4	Steam	94.3	0.0823	1.3298	647.5
S-5	Steam	33.6	0.0052	11.3775	1111.8
S-6	Steam	321.5	1.1000	3.9307	3931.6
W-1	Water	33.6	0.0052	12.7073	15.0
W-2	Water	33.6	0.5390	12.7073	21.9
W-3	Water	89.9	0.3290	12.7073	386.4
W-4	Water	104.3	0.1180	16.9242	725.0
W-5	Water	105.2	5.8800	16.9242	835.3
W-6	Water	110.0	1.1000	3.9307	193.8
E-1	Electricity	\	\	\	12483.7

and then more expensive dryer. The exergetic efficiency of the heat exchange process of the boiler $\left(\eta_{bh}^{ex}\right)$ could also be increased.

8.5.4 Cost Effective Analysis

The integration of dryers can significantly increase the efficiency of biomass-fired power plants as discussed above, but the cost effectiveness of dryers' integration

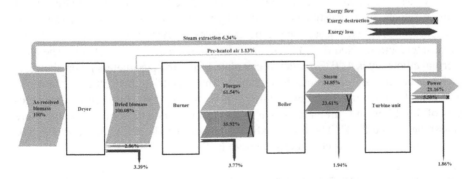

FIGURE 8.14 Sankey diagram of exergetic analysis of the PPDS.

is questionable. The thermodynamic performance and economic performance can stand contrarily or coordinately. For example, a lower drying temperature means less irreversibility during the drying process; however, a smaller heat transfer temperature difference will lead to a bigger and then more expensive dryer. Cost effective analysis is a complex problem. The thermo-economic or exergo-economic optimizations are very important for the application of dryers in biomass-fired power plants. The parameters of the drying system (heat source, thermal efficiency, etc.) influence the efficiency improvement profit by the integration of dryers. The cost of a dryer is highly technology-specific;[18] it can range from \$22 to \$796 per kg h^{-1} of water evaporated. Under the thermodynamic parameters in Section 8.5, the authors calculate the levelized cost of electricity (LCOE), when the cost of a dryer is assumed to be \$80 per kg h^{-1} of water evaporated. Results indicate that the integration of flue-gas dryer or steam dryer does not decrease the LCOE relative to the LCOE of CPP under the benchmark condition. The LCOE increases from \$0.0860 (kW h)$^{-1}$ for CPP to \$0.0863 (kW h)$^{-1}$ and \$0.0884 (kW h)$^{-1}$ when flue-gas dryer and steam dryer are integrated, respectively.

8.6 INFLUENCE OF LIGNITE DRYING ON PERFORMANCE OF POWER PLANTS*

For the lignite drying methods mentioned above, a reference case is examined and analyzed in order to predict their applications in the power plant in this section. The PPDF and PPDS are compared quantitatively: the improvements in plant thermal efficiency with different drying methods are calculated and compared.

8.6.1 REFERENCE CASE

The reference case analyzed here is a 600MW supercritical condensing power plant, which is a CPP with lignite direct-firing. Parameters of the heat regenerative system under turbine heat acceptance (THA) condition are shown in Table 8.12.

The steam turbine works in a supercritical condition. The live steam temperature and pressure are 24.2 MPa and 566°C respectively. The efficiency of the steam turbine unit is 47.81%, the mechanical efficiency is 99.5%, and the generator efficiency is 99%. The raw fuel under consideration is Yimin lignite (mined in Eastern Inner Mongolia of China). The chemical constituents of Yimin lignite are shown in Table 8.2. The boiler efficiency is 91%. The heat-supply pipe efficiency is 99%. Furthermore, the boiler exhaust temperature and the air excess ratio are 148°C and 1.2 respectively. The plant thermal efficiency is 43.07%, which is calculated from the parameters given above.

In the benchmark condition, the moisture content is dried to 19.5%, i.e., the moisture mass from drying 1 kg raw lignite (λ) is 0.2484 kg/kg. The chemical constituents of dried Yimin lignite are shown in Table 8.13. The environment temperature is assumed to be 25°C. The boiler exhaust temperature is 131°C in the PLPP. The thermal efficiency of dryers is considered to be 98%.

* Some contents in this section has been published in authors' paper (https://doi.org/10.1016/j.energy.2 012.10.026).

TABLE 8.12
Parameters of the Heat Regenerative System Under THA Condition

Heater No.	Steam Extraction			Heater Enthalpy (kJ/kg)		
	Enthalpy (kJ/kg)	Temperature (°C)	Pressure (Mpa)	Drain Water	Feed Water	Outlet Water
1	2486.4	62.5	0.02231	163.3	140	246.5
2	2622.2	85.9	0.05996	268.4	246.5	344
3	2739.4	132.5	0.1228	366.1	344	425.5
4	2979.2	257.5	0.4232	447.7	425.5	594.4
5	3199.4	369.8	1.064	\	594.4	764.9
6	3410.3	474.8	2.149	811.5	801.6	927.27
7	2980	309.4	4.266	943.9	927.27	1098.6
8	3059.6	355	6.012	1124.5	1098.6	1207.6

TABLE 8.13
Compositions of Dried Yimin Lignite

C (wt%)	H (wt%)	S (wt%)	O (wt%)	N (wt%)	Ash (wt%)	H_2O (wt%)	LHV (MJ/kg)
46.02	2.70	0.19	15.04	0.47	16.09	19.50	16.52

8.6.2 DRYING WITH BOILER FLUE-GAS

The temperature of the dryer exhaust is t_{de}, and the inlet temperature is t_{fg}. The flue-gas temperatures needed for different kinds of dryers are shown in Table 8.14. At different inlet temperatures, the flue-gas is extracted from different places in the boiler flue, or mixed with two different extractions to achieve the temperature needed. The heat transfer components in the boiler furnace have to be designed and arranged anew. However, as the aim of this paper is to give a simple way to analyze the influence of the flue-gas temperature, the detailed arrangement for flue-gas extraction is not considered. Using the dryer energy balance model, the mass flow rate of the flue-gas is calculated, the results of which are also predicted. For the calculations, the input parameters and the results are given in Table 8.14, in which "Flue-gas ratio" is the flue-gas ratio of the total flue-gas extracted to the dryer, and the "Efficiency improvement" is the plant thermal improvement of the PPDF compared with the CPP.

From the results listed in Table 8.14, we reach the following conclusions:

1. The plant thermal efficiency improvement is about 0.45 to 1.55%. In addition, there will undoubtedly be considerable environmental and economic benefits for power plants.

2. The heat capacity of the flue-gas is low, and the heat needed for raw lignite drying (moisture evaporating) is enormous. So if the inlet temperature of the dryer is not high enough, the flue-gas ratio is very high. There will undoubtedly be a great influence on the heat transfer components. As a result, a high inlet temperature of dryer must be designed for the FPLPP in CPP revamp.

3. When a lower temperature of dryer exhaust is used, greater improvement can be achieved for the plant thermal efficiency. With a lower temperature of dryer exhaust, more heat is used for lignite drying. Therefore, dryers with possible low inlet and outlet temperatures of the flue-gas dryer are preferred in a novel PPDF design.

8.6.3 STEAM DRYING

Conventional steam dryers include rotary-tube dryers and WTA dryers using extracted steam from steam turbine as a heating source. The steam at different pressures (corresponding to different saturated temperatures) can be used when the drying atmosphere pressure is varied. The performance of steam dryers is related to the degree of raw lignite drying, pressure and temperature of the extracted steam, and the dryer thermal efficiency.

The PPDS with steam extracted at different pressures (No. 3 to No. 6 heaters) used as the heat source are calculated respectively. In the calculation process, the condensate is sent to the deaerator. The results are shown in Table 8.15, where the "Extraction No." is the steam extraction number, i.e., heater No. in the regenerative system; the "Steam extraction ratio" is the steam ratio of the total steam extracted to the dryer.

TABLE 8.14
Initial Parameters[15] and Calculation Results

Dryer Type	Heating Medium Temperature (°C)		Flue-Gas Ratio (%)	Efficiency Improvement (%)	Drying Rate (kg/kg)
	Inlet	Outlet			
Rotary dryers	750	120	26.64	0.45	0.2484
Chamber dryers with stirrers	700	110	28.03	0.53	0.2484
Pneumatic dryers	600	100	32.58	0.64	0.2484
Fluid-bed dryers	450	75	41.88	0.97	0.2484
Spouted fluid bed dryers	200	80	100.00	1.55	0.1818
Vibratory dryers	300	55	62.29	1.50	0.2484
Shaft dryers	270	65	75.51	1.54	0.2484
Dryer with moving bed	175	80	100.00	1.55	0.1383

From the results listed in Table 8.15, the following conclusions could be drawn:

1. For different steam extraction numbers, the steam extraction ratio is about 6%, which is much lower than boiler flue-gas drying. The heat capacity of steam extraction is much higher than the flue-gas, which contains both sensible heat and latent heat, so lignite can be dried to a high degree with a relatively low steam extraction mass flow rate.
2. The lower the energy grade (pressure and temperature) steam extraction used as the heat source, the more plant thermal efficiency is improved. When No. 3 steam extraction was used, the plant thermal efficiency was improved by 2.19–45.26%.

The improvement potential of plant thermal efficiency for SPLPP is normally higher than that of the PPDF. The main factor is that low-grade energy (energy mostly discharged from the cooling tower) is used to dry lignite in the PPDS.

8.6.4 Cost Effectiveness Analysis

The cost effectiveness of dryers' integration is the most important indicator for the decision makers. The plant efficiency of a lignite-fired power plant could be significantly increased by drying, which would help the power plant save some fuel with constant power output or generate more electricity with constant fuel input. Many indicators could be used to analyze cost effectiveness, including LCOE, pay-back period, economic benefit, etc. As discussed in section 8.5.4, it is very difficult for drying technologies to decrease the LCOE, because the investment of conventional power plants is low and low-rank coal is very cheap. But when the input fuel is assumed as constant, a considerable economic benefit could be obtained. Xu et al.[19] evaluated the economic performance of a 1000 MW bituminous coal-fired power plant incorporating low temperature drying. Their results showed that the annual income from electricity sales significantly increased by $3.12 M because of the integration of dryers and, as a result, the additional net economic benefit of the power plant is as high as $1.91 M per year.

TABLE 8.15
Calculation Results for Steam Dryer

| Extraction No. | Parameters of Steam Extraction | | | Steam Extraction Ratio (%) | Plant Thermal Efficiency (%) | Efficiency Improvement (%) |
	Enthalpy (kJ/kg)	Pressure (MPa)	Temperature (°C)			
3	2739.4	0.1228	132.5	6.1	45.26	2.19
4	2979.2	0.4232	257.5	5.93	44.91	1.84
5	3199.4	1.064	369.8	5.78	44.6	1.53
6	3410.3	2.149	474.8	5.6	44.32	1.25

8.7   SUMMARY AND OUTLOOK

Energetic and exergetic analyses are very important for the selection, design, and optimization of drying technologies for drying of wet solid fuels which are used in power plants. Therefore, the models of energetic and exergetic analyses for coal and biomass drying were introduced. Because the integration of dryers is very attractive for power plants, models to analyze the energetic and exergetic performances of power plants integrated with dryers were also introduced in this chapter. Then, case studies were carried out on power plants with solid fuel dryers.

As the heat consumption rate could not reflect the irreversibilities of the drying process, exergy analysis on the dryer provides useful information which can identify areas for plant improvements. Case studies on drying technologies reveal that irreversibilities of drying processes vary greatly. Dryers using low temperature heat as dryer heat source consume less exergy in the drying process.

When low-grade energy is used to dry solid fuel, an integration of the dryer with the power plant can significantly decrease the exergy consumption rate of the drying process and improve the plant overall efficiency. Both the flue-gas extracted from the boiler and steam extracted from steam turbines could be used as heat source for lignite drying in a lignite-fired power plant. Case studies on influences of drying on biomass- and lignite-fired power plants reveal that the exergy consumption in combustion process significantly decreases because of fuel drying. Moreover, the energy loss with boiler exhaust decreases, and, as a result, the efficiency of power plants increases.

Based on the energetic and exergetic analyses of coal and biomass drying, more research should be carried out in the future to minimize exergy consumption during drying and promote applications of dryers:

1. Enhancing the heat transfer in the dryer. Low temperature heat source could be used in the dryer when the heat transfer between the feedstock and the heat source is enhanced. Using lower temperature drying will also be beneficial for the safety of fuel drying process and decrease the exergy consumption rate.
2. Decreasing the investment and land occupation of dryers. The fuel consumption rate in large scale power plants is enormous; therefore, it is important to consider the investment and land occupation of dryers, which are also the difficulties for the applications of dryers in power plants.
3. Improving the off-design performances of dryers. The dryers often work on off-design conditions, especially when they are integrated with power plants. The stabilization of dried fuel is very important for the operation of power plants.

ACKNOWLEDGMENTS

The authors' work was supported by the National Basic Research Program of China (973 Program, Grant Number 2015CB251504) and the National Natural Science Foundation of China (Grant Number 51436006).

REFERENCES

1. Vassilev, S. V.; Baxter, D.; Andersen, L. K.; Vassileva, C. G. "An overview of the chemical composition of biomass". *Fuel* 2010, 89(5), 913–933.
2. Liu, M.; Li, G.; Han, X. Q.; Qin, Y. Z.; Zhai, M. X.; Yan, J. J. "Energy and exergy analyses of a lignite-fired power plant integrated with a steam dryer at rated and partial loads". *Drying Technology* 2016, 35(2), 203–217.
3. Bejan, A. *Advanced Engineering Thermodynamics*. John Wiley & Sons: New Jersey, 2016.
4. Han, X.; Yan, J.; Karellas, S.; Liu, M.; Kakaras, E.; Xiao, F. "Water extraction from high moisture lignite by means of efficient integration of waste heat and water recovery technologies with flue gas pre-drying system". *Applied Thermal Engineering* 2017, 110(5), 442–456.
5. Liu, M.; Li, G.; Han, X.; Qin, Y.; Zhai, M.; Yan, J. "Energy and exergy analyses of a lignite-fired power plant integrated with a steam dryer at rated and partial loads". *Drying Technology* 2017, 35(2), 203–217.
6. Xu, C.; Xin, T.; Xu, G.; Li, X.; Liu, W.; Yang, Y. "Thermodynamic analysis of a novel solar-hybrid system for low-rank coal upgrading and power generation". *Energy* 2017, 141, 1737–1749.
7. Ma, Y.; Zhang, H.; Yuan, Y.; Wang, Z. "Optimization of a lignite-fired open pulverizing system boiler process based on variations in the drying agent composition". *Energy* 2015, 81, 304–316.
8. Avagianos, I.; Atsonios, K.; Nikolopoulos, N.; Grammelis, P.; Polonidis, N.; Papapavlou, C.; Kakaras, E. "Predictive method for low load off-design operation of a lignite fired power plant". *Fuel* 2017, 209, 685–693.
9. Kakaras, E.; Ahladas, P.; Syrmopoulos, S. "Computer simulation studies for the integration of an external dryer into a Greek lignite-fired power plant". *Fuel* 2002, 81(5SI), 583–593.
10. Luk, H. T.; Lam, T. Y. G.; Oyedun, A. O.; Gebreegziabher, T.; Hui, C. W. "Drying of biomass for power generation: A case study on power generation from empty fruit bunch". *Energy* 2013, 63, 205–215.
11. Liu, Y.; Aziz, M.; Fushimi, C.; Kansha, Y.; Mochidzuki, K.; Kaneko, S.; Tsutsumi, A.; Yokohama, K.; Myoyo, K.; Oura, K. "Exergy analysis of biomass drying based on self-heat recuperation technology and its application to industry: A simulation and experimental study". *Industrial & Engineering Chemistry Research* 2012, 51(30), 9997–10007.
12. Brammer, J. G.; Bridgwater, A. V. "Drying technologies for an integrated gasification bio-energy plant". *Renewable and Sustainable Energy Reviews* 1999, 3(4), 243–289.
13. Liu, M.; Zhang, X.; Han, X.; Li, G.; Yan, J. "Using pre-drying technology to improve the exergetic efficiency of bioenergy utilization process with combustion: A case study of a power plant". *Applied Thermal Engineering* 2017, 127, 1416–1426.
14. Kakaras, E.; Ahladas, P.; Syrmopoulos, S. "Computer simulation studies for the integration of an external dryer into a Greek lignite-fired power plant". *Fuel* 2002, 81(5), 583–593.
15. Mujumdar, A. S. *Handbook of Industrial Drying*. CRC Press: Boca Raton, 2014.
16. Kawabata, M.; Kurata, O.; Iki, N.; Tsutsumi, A.; Furutani, H. "System modeling of exergy recuperated IGCC system with pre- and post-combustion CO_2 capture". *Applied Thermal Engineering* 2013, 54(1), 310–318.
17. Nomura, R.; Iki, N.; Kurata, O.; Kawabata, M.; Tsutsumi, A.; Koda, E.; Furutani, H. "System analysis of IGFC with exergy recuperation utilizing low-grade coal". *Proceedings of the ASME Turbo Expo* 2011, 2011, 4, 243–251.

18. Amos, W. A. "Report on Biomass Drying Technology". National Renewable Energy Lab., Golden, CO (US), 1999.
19. Xu, C., Xu, G., Zhu, M., Dong, W., Zhang, Y., Yang, Y., Zhang, D. "Thermodynamic analysis and economic evaluation of a 1000 MW bituminous coal-fired power plant incorporating low temperature pre-drying (LTPD)". *Applied Thermal Engineering* 2016, 96, 613–622.

Index

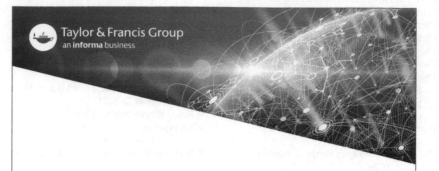